Ram Air
Ram Air

OFFICIAL PACE CAR
78th INDIANAPOLIS 500 MAY 29, 1994
INDIANAPOLIS
500

Modern American Muscle

Alan Paradise

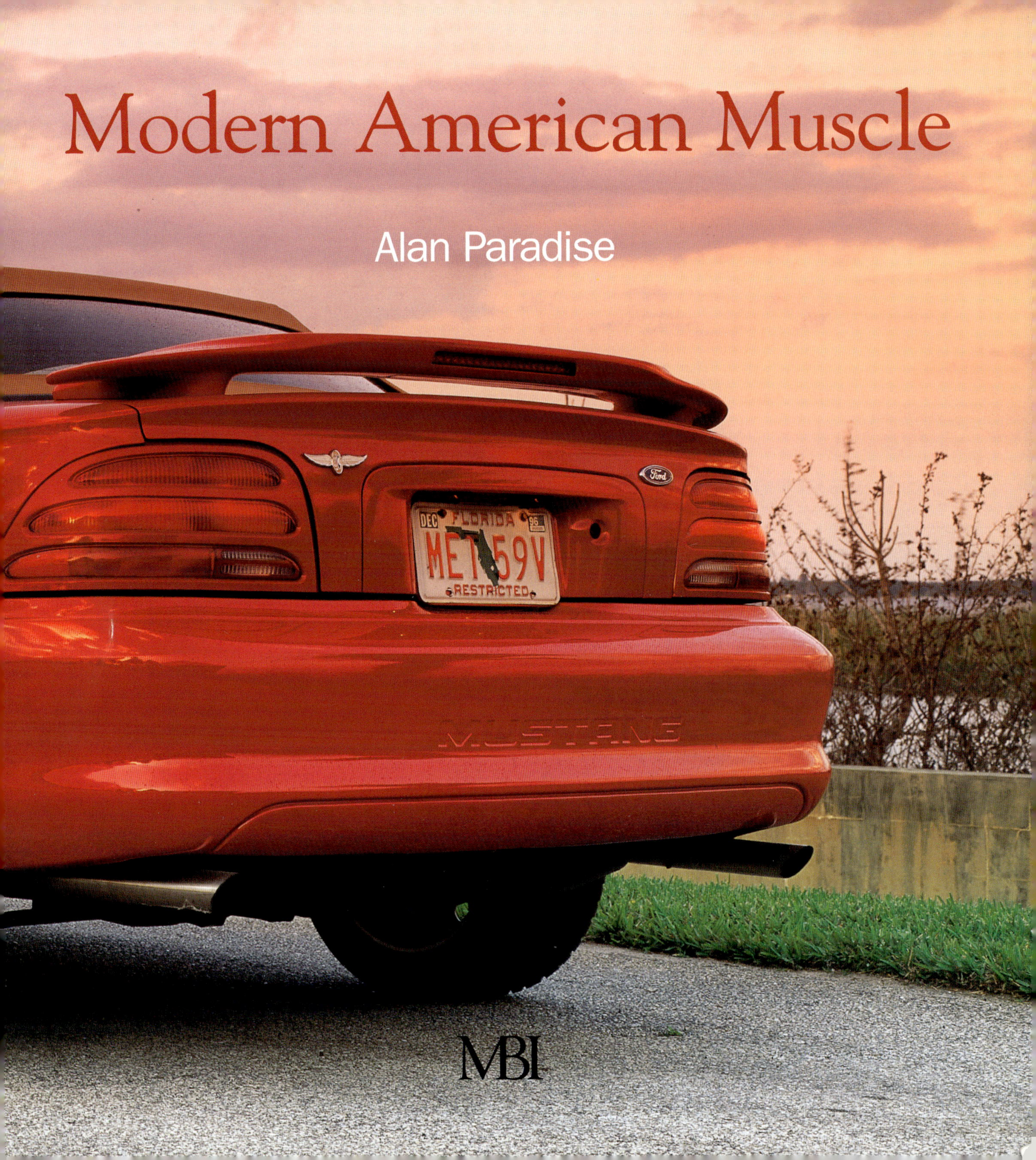

MBI

Acknowledgments

The author would like to thank everyone at DaimlerChrysler, Ford, and General Motors who assisted in the research for this book. Chrysler's Julie Butkus and GM's Jeff Youngs and Pattie Garcia stand out for their efforts. My gratitude also goes out to SLP Engineering. It is gratifying to know that performance car enthusiasts still exist within the corporate walls of the domestic automakers.

Special thanks to the owners of the cars and trucks featured in this book—for sharing their vehicles and their enthusiasm for modern American musclecars and trucks.

This edition published in 2003 by MBI, an imprint of MBI Publishing Company, Galtier Plaza, Suite 200, 380 Jackson Street, St. Paul, MN 55101-3885 USA

First published in 2001 by MBI Publishing Company

ISBN 0-7603-1830-1

Front cover: At home on the road or at the track, the Viper RT/10 Roadster lives up to the Mopar legacy of powerful R/T models from the original musclecar era. While red was the only color originally available, this 1995 roadster is one of 307 that year to sport the Emerald Green exterior that first became available in 1994. Side curtains and a snap-on top are more reminiscent of 1950s British sports cars than 1960s American iron.

Front banner: This 1997 SS Coupe features modifications by SLP Engineering, which were ordered through the Cheverolet dealer, that included a special hood with functioning scoop and forced induction, revised suspension, and 17x9-inch Corvette ZR-1 wheels.

On the frontispiece: The twin hood nostrils let the world know that this is the heavy-breathing version of the Trans Am; capable of obtaining 14-second quarter-miles and 5.7-second 0-to-60 sprints.

On the title page: Mustang celebrated its 30th birthday by pacing the 1994 Indy 500. The pace car was a red SVT Cobra convertible. Actually, there were three pace cars with A.J. Foyt and Parnelli Jones as guest drivers. These were the first-ever Cobra ragtops, and 1,000 replicas were produced and sold.

On the back cover: This Cobra R looks ready to do high-speed wind sprints with 18x9.5-inch wheels and aerodynamic acoutrements. The side exhaust not only sounds and looks cool, but it is also necessary to avoid racing fuel cell and independant rear suspension setup. Stiffer Eibach springs lower the R model 1.5 inches in front and 1 inch in the rear versus the standard Cobra.

Printed in China

Contents

Introduction

It is a little bit of a misnomer to call this book *Modern American Muscle.* The vehicles covered here are evolutions of the factory hot rods that originated the term "musclecar," but this generation of performance machinery relies on brains, not brawn, to generate driving thrills and excitement. The original musclecars were products peculiar to the 1960s. Musclecars were tire-screeching, pavement-pounding, street-scorching, window-rattling, V-8–powered deviants that rolled off United Auto Workers (UAW) assembly lines alongside family sedans and station wagons for delivery to your friendly neighborhood General Motors (GM), Ford, or Chrysler dealer. The same dealers and manufacturers, accompanied by a musical score composed by rock groups including the Beach Boys, Jan and Dean, and Ronnie and the Daytonas, slugged it out on the drag strip as well as the showroom in hopes of capturing what marketers tabbed the "youth market." Things eventually change, and concerns about safety, emissions, and petroleum resources ended the musclecar era. The youth market grew up to be baby boomers. America became a kinder, gentler P.C. society—political correctness and personal computers.

Back in the "Swingin' '60s," the hot rodders' criterion was that there is no substitute for cubic inches. That was before microprocessors started running our lives and cars. As we headed into the 1990s, computer-controlled engine management systems that optimized the balance between fuel economy and emissions became standard equipment, and a new generation of technoid hot rodders discovered that these systems could also be programmed for high performance. Synapses displaced sinews as a means to provide scintillating performance for a new group of modern-day factory hot rods. these vehicles sparked the high-performance memories of baby boomers, while a new generation, spurred on by virtual-reality thrill rides instead of rock bands, became addicted to high-horsepower thrills.

The original Big Three domestic automakers are still the primary source for these modern American musclecars, but like the customers they are trying to reach with these specialty vehicles, the automakers have also undergone dramatic changes. Pickup trucks have replaced family sedans as the bread-and-butter vehicles for GM and Ford. This is reflected in the SS454 and Ford Lightning that are included in our collection of high-performance machinery. The once distinct flavors of the various divisions of GM that provided a banquet of drag-strip delights such as the GTO, 442, GS455, and Chevys of all shapes and sizes, have turned to mush as the company struggles to regain direction and market share, thanks to a disastrous corporate branding strategy that achieved the opposite effect. The Corvette still rules the roost, but none of the other GM performance cars survived the 1980s except for the Camaro and Firebird, which are destined to fade into musclecar memories in 2002.

The Mustang is a high-performance survivor at Ford where world cars are more in focus (pun intended) today than brutish American musclecars. Speaking of international, Chrysler is now part of the German company, DaimlerChrysler. Where and what will happen to it in the future is anyone's guess, but remarkably during the 1990s, it was Chrysler who turned out the cars—the Viper and Prowler—that were the closest embodiment of the American factory hot rods. Like Ford and GM, Chrysler also played to the growing truck trade with a couple of hot pickups.

As you flip through the following pages of the latest generation of musclecars and trucks, keep in mind that although the definition of a musclecar may change with the times, the one thing that never varies is the aim to separate their owners from the pack, whether you're standing still or tromping on the gas. Here's to musclecars of the past, present, and future.

—Patrick C. Paternie

Saleen took the already potent Mustang GT and turned it into an extraordinary sports car. The Saleen Mustang is arguably the most desirable tuner special Mustang available today. This 1997 S281 features an all-aluminum, supercharged 281-ci V-8 that produces about 350 horsepower and can run from 0 to 60 in about 5 seconds.

5.0

1

FORD'S FLYING CIRCUS OF PONIES AND PICKUPS

The resurgence of the Mustang GT during the 1980s as a favorite of hot rodders and sports car enthusiasts provided the momentum for Ford to increase its performance image as it headed into the 1990s. The pony car remained Dearborn's weapon of choice, but selling performance in the 1990s required a more diverse arsenal since the musclecar battles that had been fought at the drag strip during the 1960s had now spread to parking-lot autocrosses and road courses.

Besides duking it out with traditional rivals GM and Chrysler, there was now significant performance competition from Europe and Japan. Trick pony cars, that could only wobble around corners, were scorned and replaced with well-balanced cars

The 1993 Mustang 5.0 GT convertible represents the last year of the Fox-platform cars that, beginning in 1982 with the initial 5.0 GT, rekindled the musclecar spirit in new domestic vehicles.

Besides updating the Mustang's exterior from a boxy 1980s' look to a sleeker, more windswept appearance, the SN95 chassis that debuted in 1994 featured separate coupe and convertible configurations. Convertibles were an afterthought in the previous-generation Fox-platform cars, and were created by chopping off the roof.

capable of stopping, cornering, and accelerating. Mustang fans, joined by patriotic musclecar enthusiasts, led a successful campaign that convinced Ford to keep the Mustang faithful to its red, white, and blue V-8powered, rear-drive roots.

At the same time, another traditional American vehicle was also becoming a factor in the marketplace. The pickup had migrated from farms to the cities and suburbs. As with any popular vehicle, there were owners who wanted a truck that was faster and handled better, hence the creation of the F-150 Lightning.

In 1991, Ford answered these challenges of strength and power by establishing its Special Vehicle Team. The SVT is an in-house band of performance enthusiasts who have developed vehicles including the Mustang Cobra and the F-150 Lightning, so the Ford Motor Company can flex some modern American muscle.

Mustang celebrated its 30th birthday by pacing the 1994 Indy 500. The pace car was a red SVT Cobra convertible. Actually, there were three pace cars with A. J. Foyt and Parnelli Jones as guest drivers. These were the first-ever Cobra ragtops, and 1,000 replicas were produced and sold.

Mustang GT

When Ford announced that the "Boss was Back" with its 1982 5.0-liter Mustang GT, this signaled the re-emergence of new-car performance after a decade of struggles with emissions and fuel economy requirements. It also rekindled the art of hot rodding cars off the showroom floor. An entire aftermarket industry developed around the Mustang GT's 5.0-liter engine

and the Fox-platform chassis. While it may seem blasphemous to the "bow-tie" supporters, many people consider these cars to be the 1955 Chevys of the 1980s in terms of their popularity and adaptability to performance modifications.

Adaptability is one thing, but flexibility is another, especially when it applies to the nature of a chassis that dated back to 1979. By 1993, the Fox-platform Mustang was 15 years old and Chevrolet had an improved Camaro to tempt pony car shoppers. It was time for a new Mustang. Despite a scary moment or two in the late 1980s when it seemed that the next Mustang might evolve into a four-cylinder, front-wheel-driver made in Japan, Ford eventually listened to the protests of Mustang lovers and its own good sense. The new 1994 Mustang followed the traditional

Above: *The most distinctive external feature of the 1993 Cobra R was the black trim on the special 17-inch alloy wheels. These wheels resurfaced in 1994 as an option on the Mustang GT. Ford produced and sold 107 of the $25,692 Cobra R models.*

Left: *The Fox platform went out with a bang in the summer of 1993 when SVT made a limited run of 107 Cobra R models that were intended for SCCA and IMSA racing. The changes to a regular Cobra included a stiffer suspension setup with Koni shocks and Eibach springs, and the removal of the air conditioning, power windows, power locks, sound insulation, and back seat in order to shed 150 pounds.*

A revised intake system and more-free-flowing heads boosted the 1993 5.0-liter Cobra R engine's power output to 235 horsepower, 30 more than in the GT. The more potent engine did not run out of breath above 4,000 rpm as did the stock GT unit.

front-engine/rear-drive format. The boxy 1980s look was swept aside by wind-sculpted, swoopy new bodywork. Similar to the original Mustang, the new car had a chrome running horse emblem in the center of the grille and a "C" scoop on its flanks just ahead of the rear wheel. Inside, a retro-styled twin-pod dash was a reminder of the first models.

The new chassis was 56 percent stiffer in bending and 44 percent stiffer in torsion than the Fox platform. A big plus for those who enjoy top-down motoring was that the new Mustang also had a specially designed convertible chassis. The previous-generation Mustang ragtops were chopped-top conversions of hardtop cars.

The stiffer chassis combined with tweaks to the MacPherson strut front suspension and live rear axle carried over from the old car to improve the handling of the new Mustang GT. The standard GT wheels were similar to those of the 1991–1993 models, with 16x7-inch alloys carrying 225/55 ZR16 tires. The big news was an optional 17-inch wheel setup that had 245/45 ZR17 Goodyear Eagle GSC tires. This gave Mustang owners a one-up on Camaro and Firebird drivers who had no such option at the time. Ford also atoned for past performance sins by finally making four-wheel disc brakes standard on the GT.

The 5.0-liter V-8 carried over to the new GT, but that was not all good news. Rated at 215 horsepower and mated to a Borg-Warner T-5 five-speed manual transmission (a four-speed automatic was optional), the GT came up one gear and 60 horsepower

2000 FORD MUSTANG GT SPECIFICATIONS

Body/Chassis	Unit steel
Engine	4.6-liter SOHC V-8, two valves/cylinder, iron block, alloy heads, electronic sequential fuel injection
Power	260 horsepower @ 5,250 rpm 302 ft-lbs torque @ 4,600 rpm
Transmission	Five-speed manual
Suspension	Front: MacPherson struts, lower A-arms, coil springs, stabilizer bar Rear: Live axle, angled upper and lower trailing arms, coil springs, tube shocks, stabilizer bar
Brakes	Four-wheel vented discs with ABS
Wheels/Tires	16x7.5-inch cast-aluminum alloy; 225/55 HR16 (17-inch optional)
Wheelbase	101.3 inches
Length	183.2 inches
Curb Weight	3,410 pounds
EPA Fuel Economy, City/Highway	17/24 miles per gallon

The interior of the 1993 Cobra R was still plush for a racecar despite the weight-reduction plan that deleted many creature comforts including the air conditioning and stereo. The lack of racing seats was a problem not addressed until the 2000 model.

SVT got even more serious with the 1995 Cobra R. Although 250 models were produced, a racing license was necessary to purchase the car as well as a check for $35,499—plus $2,100 in taxes—to discourage speculators from crowding out the intended customers.

short of the new Camaro, which had a 275-horsepower V-8 and six-speed transmission. *Road & Track* magazine recorded a 0-to-60-mile-per-hour time of 6.7 seconds, and a 15.2-second quarter-mile in its test of the new 1994 GT. This was 0.4 seconds slower than the Camaro in both categories.

Ford sought to improve things in 1996 when it replaced the pushrod 5.0-liter V-8 with a high-tech modular 4.6-liter single-overhead-cam (SOHC) unit with alloy heads. This engine was also rated at 215 horsepower, although a 305-horsepower, all-alloy, dual-overhead cam (DOHC), four-valve-per-cylinder version was installed in the 1996 SVT Cobra Mustang. The new engine was a technical tour de force, much smoother and higher revving, but when it came time to put the pedal to the metal at the drag strip, the new engine was a step backward. *Road & Track* numbers showed the 1996 GT to be 0.1 second slower both from 0 to 60 miles per hour and in the quarter-mile than its 1994 5.0 counterpart.

To celebrate the Mustang's 35th anniversary, Ford made a number of styling and performance changes to

The 17x9-inch wheels of the 1995 Cobra R were the largest ever offered up to then on a Mustang. As a result, they have become a cult item among Mustang aficionados who are searching for that extra edge in handling and style.

SVT followed the credo that there is no substitute for cubic inches when they designed the 1995 Cobra R to do battle with road-racing Camaros. Out went the 5.0-liter V-8 and in went a 5.8-liter (351 cubic inches) Windsor V-8. A trick camshaft, alloy pistons, forged connecting rods, and special heads and intake helped wring 300 horsepower and 365 ft-lb of torque out of the old warhorse.

the 1999 Mustang. The appearance changes tried to capture a retro look reminiscent of the original car. Not everyone appreciated the restyling, but performance fans had to be happy with the changes under the revised hood. The 4.6 liter gained 35 horsepower for a total of 260 horsepower at 5,250 rpm. This was still 45 horsepower shy of a comparable Z28 Camaro. According to *Motor Trend*'s testers, who tend to be quicker and less sensitive to drivetrains and tires than those of *Road & Track*, the anniversary Mustang went from 0 to 60 miles per hour in 5.4 seconds. *MT*'s time for the Z28 is a flat 5 seconds. *Consumer Guide* recorded a time of 6.5 seconds. A real-world number would probably be in the low 6-second range.

The Mustang GT may not be the performance leader it once was. The Cobra is now Ford's top dog—make that pony—but the Mustang GT has matured into a capable sporty car worthy of wearing the grand touring initials.

SVT Mustang Cobra

Ford's Special Vehicle Team (SVT) was formed in late 1991 to create limited-production specialty vehicles. Think of them as very sophisticated corporate hot rodders. The people who design and build these special vehicles within the SVT are the Special Vehicle Engineering (SVE) group.

The first vehicles to emerge from the SVT garage were 1993 high-performance variants of the F-150 pickup and the Mustang. The SVT version of the Mustang reintroduced the Cobra name as a potent icon of Ford performance. It was also the last hurrah for the Fox platform, which would be replaced by a new Mustang the following year.

From the outside, the 1993 SVT Cobra featured subtle but handsome modifications to distinguish it from the GT. The most noticeable changes were a tiny grille opening in the nose featuring a chrome running horse badge and new taillights. Exterior colors were limited to red, black, and teal. The Cobra also featured 17-inch wheels for the first time on a Mustang. Measuring 17x7.5 inches, the wheels were fitted with 245/45 ZR17 tires. The GT's 5.0-liter V-8 resided under the hood, but any resemblance to the GT was purely coincidental. The SVE had fitted a new upper and lower intake manifold as well as a bigger throttle body and mass air sensor. Injectors also went up in size and were fed by a high-flow fuel pump. Free-breathing heads featuring larger intake and exhaust ports were installed. Corresponding changes were made to the engine management electronics and cam profile. The bottom line added up to 235 horsepower. The five-speed manual transmission was revised to handle the new power output, and four-wheel disc brakes were also part of the Cobra package.

The Cobra was more than just a one-trick pony as the SVE went to work on the suspension and utilized the talented backsides of Jackie Stewart and Bob Bondurant to help calibrate the optimum ride and handling setup.

Motor Trend tested a 1993 Cobra and came up with a 0-to-60 time of 6.2 seconds. Elapsed time for the quarter-mile was 14.4 seconds at 97.4 miles per hour. These times were quick enough for the SVT to sell 4,993 Cobras against its target of 5,000 units.

A new Mustang debuted in 1994, and the SVT was ready with a new Cobra model. Engine output was increased to 240 horsepower, and larger brake discs (13 inches up front, 11.65 inches in back) were fitted. Revisions were also made to the suspension, which primarily decreased spring rates and fitted smaller stabilizer bars from those of the 1993 Cobra. New 17x8.0-inch wheels were also part of the package. Subtlety was again the guiding force in exterior differentiation from the GT. Larger, round fog lamps replaced the rectangular units of the GT, and a slightly altered rear spoiler was fitted. Inside, white gauge faces made their debut as an SVT trademark.

To celebrate the Mustang's 30th birthday, a red Cobra convertible was selected to pace the 1994 Indy 500. The SVT produced 1,000 Pace Car Edition Cobras to commemorate the event. These were the first convertible Cobras ever offered. In 1995, a convertible model, available only in black, was added to the otherwise unchanged Cobra lineup. Over 5,200 Cobras were sold that year.

2001 SVT MUSTANG COBRA SPECIFICATIONS

Engine	4.6-liter all alloy, DOHC V-8, four valves/cylinder, electronic sequential fuel injection
Power	320 horsepower @ 6,000 rpm (6,800 redline) 317 ft-lbs torque @ 4,750 rpm
Transmission	Tremec five-speed manual/limited-slip differential and traction control
Suspension	Front: MacPherson struts, lower A-arms, variable rate coil springs, tube shocks, stabilizer bar Rear: Independent, iron upper and aluminum lower control arms, linear rate springs, tube shocks, stabilizer bar
Brakes	Four-wheel vented discs with ABS
Wheels/Tires	17x8-inch forged-aluminum alloy; 245/45 ZR17
Wheelbase	101.3 inches
Curb Weight	3,446 pounds
EPA Fuel Economy, City/Highway	17/24 miles per gallon

And now for something completely different: the 2000 model Cobra R eschews the subtle appearance of its predecessors for an evil, hard-core racer look. Stuffing the taller 5.4-liter V-8 into the Mustang engine bay required lowering a chassis cross-member. More noticeable change was the tumescent hood, also a necessity for adequate clearance. SVT built 300 2000 Cobra R models with a price tag of $55,000.

Borrowed from the Lincoln Navigator, the Cobra R's 5.4-liter V-8 used aluminum high-flow, four-valve heads and forged-alloy pistons to bring power levels up to 385 horsepower at 5,700 rpm, with 385 foot-pounds of torque at 4,500 rpm.

2000 MUSTANG COBRA R SPECIFICATIONS

Engine	5.4-liter DOHC V-8, four valves/cylinder, iron block, alloy heads, electronic sequential fuel injection
Power	385 horsepower @ 6,250 rpm 385 ft-lbs torque @ 4,250 rpm
Transmission	Tremec six-speed manual
Suspension	Front: MacPherson struts, lower A-arms, coil springs, stabilizer bar Rear: Independent, upper and lower arms, coil springs, tube shocks, stabilizer bar
Brakes	Four-wheel vented discs with ABS
Wheels/Tires	18x9.5-inch cast-aluminum alloy; 265/40 ZR18
Wheelbase	101.3 inches
Curb Weight	3,590 pounds
EPA Fuel Economy, City/highway	13/18 miles per gallon

Cobra R looks ready to do high-speed wind sprints with 18x9.5-inch wheels and aerodynamic accoutrements. The side exhaust not only sounds and looks cool, but it is also necessary to avoid racing fuel cell and independent rear suspension setup. Stiffer Eibach springs lower the R model 1.5 inches in front and 1 inch in the rear versus the standard Cobra.

Things became even more exciting in 1996 when Cobras received their own special engine in the form of a 305-horsepower, all-alloy, four-valve, DOHC 4.6-liter V-8. Each engine is hand-assembled by two-man teams that work in a dedicated section of Ford's Romeo, Michigan, engine plant. It takes a little over an hour for the specially trained teams to assemble an engine. In order to complete this custom approach to engine building, the assembly team adds one more personal touch by autographing a metallic decal that is applied to the passenger-side cam cover.

Other changes to 1996 Cobras included larger 3-inch exhaust tips, firmer suspension settings, and a revised hood with twin power bulges to clear the new engine's intake system. The rear wing also became a customer-delete option. A chameleon-like Mystic Clearcoat paint, which changed hues according to lighting conditions, was available on the Cobra coupe for an extra $815.

With horsepower up to 305 at 5,800 rpm, the Cobra's 0-to-60 times went down accordingly. *Road & Track* recorded 5.9 seconds in a test pitting the SVT Cobra against a 1996 305-horsepower Camaro SS. With identical horsepower ratings, the two cars turned identical times. Not surprisingly, 10,006 Cobras were sold in 1996, and 1997 sales were also at the 10,000-unit limit imposed by the SVT.

The front splitter attaches with Dzus fasteners and works to reduce front-end lift, especially important since SVT claims a 170-mile-per-hour top speed for the car.

For 1999, as with all Mustangs, the exterior of the Cobra took on a new angular appearance. The new body allowed for a 1.4-inch-wider rear track for improved handling. The handling improved even more thanks to a new independent rear suspension in place of the live axle that was still used on other Mustangs.

Besides the improved rear suspension, the big news surrounding the 1999 Cobra was a horsepower rating of 320 for the 4.6-liter engine. Unfortunately it was not good news as many owners accused Ford of letting some of those horses slip out of the barn before delivery. As a result, there were no 2000 model year Cobras produced because Ford rounded up the strays and made modifications to all the 1999s it could track down.

Ford guarantees that the 2001 Cobra will have all 320 horses as advertised. New features on the 2001 Cobra include a rear bumper embossed "Cobra" instead of "Mustang," and new suede seats embroidered with the serpent logo.

Mustang Cobra R

R stands for "racing" just as it did back in 1965 when Carroll Shelby built a limited run of 37 R-Model Shelby GT 350s. Hoping the R would also stand for a repeat of the on-track success enjoyed by Shelby's cars, Ford gave the okay in 1993 to a run of 100 Cobra R competition models to do battle against Camaros and Firebirds in the Sports Car Club of America (SCCA) World Challenge and the International Motor Sports Association (IMSA) Firestone Grand Sport Series.

Starting with an SVT Mustang Cobra, a less-is-more philosophy prevailed as air conditioning, power windows, fog lights, sound insulation, inner-fender panels, and the back seat were stripped away to shed about 150 pounds. Koni shock absorbers and struts,

As subtle as a hand grenade, the 2000 Cobra R features a high-speed aero package to ensure stability at high speeds. With a curb weight just under 3,600 pounds, the R is 160 pounds heavier than a standard Cobra, and has 65 more horsepower.

The 1993 F-150 Lightning joined the Cobra as the first models to be produced by Ford's Special Vehicle Team (SVT). Based on sales, the Lightning was the more popular of the two, and outsold the Cobra 5,276 to 4,993. A monochromatic paint scheme (available in red, white, or black), front air dam with fog lamps, and 17-inch aluminum five-spoke wheels were visual clues that this pickup was meant to haul more than hay. First-generation SVT Lightnings were made from 1993 to 1995.

A sliding rear window, black power mirrors, and tonneau cover were all options for the F-150 Lightning. Monroe Formula GP shock absorbers and 1-inch front and rear stabilizer bars improved handling while still allowing a payload of 700 pounds and a 5,000-pound trailer towing capability.

along with larger stabilizer bars and stiffer Eibach springs, replaced their more ride-compliant counterparts. Five-bolt 17-inch wheels and upgraded brakes were also part of the R modifications. An oil cooler was added along with increased cooling capacity for the engine.

Priced at $25,692, all 107 of the hand-built cars were sold before they rolled off the assembly line. Unfortunately, the R was more of a sales success than a racing success. Speculators, rather than racers, gobbled up the majority of the cars. The Rs that did hit the track were outgunned by Camaros that toted 40 more horses.

Ford was determined to correct this situation with the next R model, which came out in 1995. A 5.8-liter Windsor V-8 replaced the 5.0-liter V-8. The engine had a marine application block and sported a special cam, aluminum alloy pistons, forged-steel connecting rods, GT-40 heads and lower intake manifold, and a special-cast alloy upper intake. A beefier Tremec five-speed transmission was used to handle the extra power along with an 8.8-inch rear end. This vehicle was good enough to make 300 horsepower at 4,800 rpm and 365 foot-pounds of torque at 3,750 rpm, and still meet emissions standards as a street-legal vehicle.

The standard Cobra body underwent a similar diet plan as applied to the 1993 model: no radio, air, or back seat. Likewise, suspension changes included the installation of Koni shocks and struts as well as Eibach springs. Other modifications were a 20-gallon racing fuel cell and additional cooling for the engine and power steering. The wheels were upgraded to 17 x 9 inches. All 250 cars were white with a beige cloth interior. Priced at $35,499—plus a $2,100 gas-guzzler tax—the R also required customers to have something else besides a fat bank account. To keep speculators at bay, Ford required buyers to present a racing license.

The 1995 Cobra R was a serious racecar as evidenced by *Motor Trend*'s comparison test to a Shelby R-Model GT 350. The newer car went from 0 to 60 miles per hour in 5.2 seconds as opposed to the Shelby's time of 5.5 seconds. The R model did get its revenge at the drag strip with an elapsed time of 13.6 seconds versus 13.8 for the R.

For 2000, the latest R model is a horse of a different color, literally and figuratively. All 300 vehicles are red with charcoal interiors, but it is the philosophical as well as technical changes that are worth noting. In the engine bay, you will see a 5.4-liter V-8 resembling that of the Lincoln Navigator. The SVT took the cast-iron truck block and added special aluminum four-valve heads that increased the airflow up to 25 percent over the Cobra. Forged-aluminum pistons, billet steel connecting rods, and a large-capacity racing oil pan are also part of the package. Exhaust gases flow through stainless-steel headers to special Borla mufflers and exit through side pipes. The 5.4-liter engine makes 385 horsepower at 5,700 rpm and an equal amount of torque at 4,500 rpm.

A Tremec six-speed transmission replaced the standard five-speed transmission. Brake rotors were the same as the Cobra, with the addition of Brembo four-piston calipers on the front. Suspension modifications included Bilstein shocks

The Opal Grey interior features XLT-level trim and bucket seats with power lumbar support. The center console could be converted into a seat. Other standard equipment included air conditioning, four-speaker electronic AM/FM/cassette stereo with digital clock, power side windows and door locks, tilt wheel, cruise control, interval wipers, and driver-side air bag.

The venerable 5.8-liter V-8 had a cast-iron block with cast-iron GT-40 design cylinder heads; a two-piece, tuned-length aluminum intake manifold; tubular stainless-steel headers; and a dual stainless-steel exhaust system. With 240 horses, the 4,446-pound pickup could do 0 to 60 miles per hour in 7.6 seconds.

The second-generation SVT F-150 Lightning debuted in 1999. It represented a major step forward in handling and performance for pickup trucks. As with the first generation, the only choices for exterior colors were red, black, or white, but that is where the similarity ends. A supercharged 360-horsepower 5.4-liter V-8 lowered 0 to 60 under 6 seconds. Ford produced about 2,500 1999 Lightnings.

and stiffer Eibach springs. The R pulls 1 g of lateral acceleration thanks to 18-inch wheels and 265/40 ZR18 BFGoodrich tires.

Unlike past R models, the 2000 model retained the power windows and power door locks from the standard Mustang. Unlike its predecessors, the R weighs 160 pounds more than a Cobra. The added weight does not slow this vehicle down. *Motor Trend* recorded 0 to 60 miles per hour in 4.82 seconds and a 13.09-second quarter-mile. With numbers like that, R also stands for respect.

Ford F-150 SVT Lightning

Unless you have had your head buried under the hood of your 409 Biscayne since the 1960s, you probably have noticed that one of the most significant evolutionary changes to American high-performance machinery over the past 40 years relates to the status of the pickup truck. Back when the Beach Boys "Shut Down" was on the Top 40 playlist, you may have owned a pickup truck to tow your musclecar to the drag strip. Today, a pickup truck could be your track vehicle.

As the manufacturer of the best-selling pickups (the F-150 is the best-selling new vehicle) in the United States, Ford should have the know-how to make a pickup into a track vehicle. If you get your hands on an F-150 Lightning, you will find that this is the case. Ford's Special Vehicle Engineering (SVE) department has spent much of the last decade refining the acceleration and handling capabilities of the F-150 to the point where the 2001 version of the SVT Lightning has 380 horsepower to propel the Regular Cab short-wheelbase pickup from 0 to 60 miles per hour in 5.8 seconds, according to Ford's stopwatches. This may be a bit conservative as *Motor Trend*'s test of a 1999 Lightning with 360 horsepower yielded a 0-to-60 time of 5.5 seconds and a big grin for the test driver who acknowledged it as the champion among new vehicles for sustained smoky burnouts. Suspension improvements (including five-leaf rear springs, gas shocks, beefy sway bars, and 295/45 ZR18 Goodyear Eagle F1-GS unidirectional tires specially developed for the Lightning) make the truck just as much fun on twisty roads.

A supercharged 5.4-liter Triton V-8 powers the 1999, 2000, and 2001 F-150 Lightnings. Thanks to improvements in engine breathing and intercooler efficiency, the 2001 model has had horsepower increased to 380 at 4,750 rpm, and torque has been bumped to 450 foot-pounds at 3,250 rpm. Boost pressure is 8.0 psi for the Eaton Gen IV Roots-type

The sweeping curves of the latest generation F-150 may look a bit too gentile for the hay hauling set, but they serve as a good foundation for the aerodynamic goodies like the front spoiler and rocker panels. Ford's Special Vehicle Engineering (SVE)—the Special Vehicle Team (SVT) members who design and build Ford's elite performance vehicles—added these pieces.

Ford

SVT
46M 254

Left: *The Lightning is built on the Regular Cab, short-wheelbase chassis to save weight. The Flareside body blends well with the added rocker panel extensions and aggressive front air dam with fog lamps. Handling improvements for the 1999 Lightning include a 31-millimeter front stabilizer bar and five-leaf rear spring with 23-millimeter stabilizer bar.*

supercharger with a water-to-air intercooler. Aluminum alloy heads, forged-steel crankshaft, dual-bore throttle body, and performance-calibrated sequential electronic fuel injection are other high-performance goodies used on the truck's V-8. The powerplant exhales through cast-iron headers and tuned dual exhaust pipes with ceramic-coated tips.

A four-speed automatic transmission and husky 4.5-inch aluminum driveshaft deliver power to the rear wheels via a 3.73 rear end with limited-slip differential. Vented disc brakes are on all four wheels. Huge 18x9.5-inch, five-spoke, cast-aluminum alloy wheels with painted surfaces provide a purposeful NASCAR look in addition to a solid foundation. Other equipment includes antilock braking system (ABS); six-way power driver's seat; six-disc CD changer; power windows, locks, and mirrors; engine oil cooler; remote keyless entry; and a Class III towing package. The only option is a factory-installed soft tonneau cover. The interior treatment includes a 40/60 bench seat trimmed in a combination of Ebony textured leather and Medium Graphite cloth,

A Roots-type Eaton supercharger with intercooler steals the show under the hood of the F-150 Lightning. Running at 8.0-psi boost, the blower pushes the 5.4-liter V-8 to 360 horsepower in 1999 and 2000 trucks. For 2001, SVT engineers have found another 20 horsepower.

A high-performance interior is part of the F-150 Lightning package, which includes custom leather-covered seats featuring additional side bolsters for cornering support. White-faced SVT gauges include a 6,000-rpm tachometer and a boost gauge. The full luxury treatment includes air conditioning, and full power. For 2001, a six-CD changer is also standard issue.

and a customized instrument panel with SVT-style white-faced gauges. The exterior trim includes a black grille, custom rocker panels, and front air dam with driving lights. The color choices are limited to red, black, silver, or white.

Lightning pickups are limited-production vehicles only sold through SVT-certified Ford dealerships, which limits availability to about 10 percent of all Ford dealers. The second-generation Lightnings debuted in 1999 with 360 horsepower, and about 2,500 1999 models were produced. The 2000 model year Lightnings were almost identical to the 1999s, and production was around 4,000 units.

The first-generation F-150 SVT Lightning was produced from 1993 to 1995. These were much less sophisticated vehicles that used the old Twin I-Beam front suspension instead of the current short- and long-arm (SLA) type, 17 x 8-inch wheels, P275/60 HR17 tires, and the venerable 5.8-liter (351 cubic inches) V-8. The latter, decked out with GT-40 design cast-iron cylinder heads, a two-piece aluminum intake manifold, tubular stainless-steel exhaust headers, and multiport electronic fuel injection put out 240 horsepower at 4,200 rpm, and 340 foot-pounds of torque at 3,200 rpm. Ford quoted a 0-to-60-mile-per-hour time of 7.6 seconds and a top speed electronically limited to 110 miles per hour. By comparison, Ford has the confidence to list the latest Lightning as being capable of an unfettered 142 miles per hour. There were 5,276 Lightnings produced in 1993, 4,007 in 1994, and 2,280 in 1995. For high-performance fans, having Lightning strike twice is a good thing.

2001 FORD F-150 SVT LIGHTNING SPECIFICATIONS

Body/Chassis	Regular Cab Flareside/short wheelbase
Engine	5.4-liter SOHC V-8 with supercharger and intercooler
Power	380 horsepower @ 4,750 rpm 450 ft-lbs torque @ 3,250 rpm
Transmission	Four-speed automatic
Suspension	Front: Short/long arm, coil springs, Bilstein gas shocks, 31-millimeter stabilizer bar Rear: Solid axle, staggered Bilstein gas shocks, five-leaf springs, 23-millimeter stabilizer bar
Brakes	Four-wheel vented discs with ABS
Wheels/Tires	18x9.5-inch cast-aluminum alloy; 295/45 ZR18 unidirectional with 235/70 R16 spare
Wheelbase	119.8 inches
Length	208.0 inches
Curb Weight	4,670 pounds
Payload/Towing Capacity	800 pounds/5,000 pounds
EPA Fuel Economy, City/Highway	13/17 miles per gallon

Besides the trick dual side outlet exhaust, standard equipment for 1999 included 18-inch wheels, a 9.75-inch rear end, and massive four-wheel vented disc brakes (12.1-inch front, 13.1-inch rear).

2

MODERN MOPAR MUSCLE

The Chrysler Corporation may not have invented the musclecar, but it certainly stretched the concept to the breaking point. Hemi-powered Roadrunners and Super Bees, tall-tailed Super Birds, and 440 'Cudas were outrageous vehicles to watch, hear, or drive.

As memorable as these cars were, Chrysler cars of the 1970s and early 1980s were forgettable. The company teetered on bankruptcy, but fought back to gain financial respectability by building bread-and-butter family transportation such as the minivan.

Churning out minivans may be good business, but it is a bit boring. Fortunately, Chrysler's president at the beginning of the 1990s was Bob Lutz. Lutz encouraged the development of exciting concept cars that were conceived by a design team headed up by Tom Gale.

The first of these concept cars to be dangled before the public was the Viper RT/10. The RT recalled the designation Dodge applied to its high-performance models of the late 1960s and early 1970s

Reminiscent of the Shelby Cobra Daytona Coupe, the 1996 Viper GTS is as pretty as the roadster is ungainly. More than just a skin graft, Chrysler engineers claim that turning the Viper into a coupe required 90 percent new parts.

Leather seating is about as luxurious as things get inside the cramped confines of this 1995 Viper's otherwise Spartan interior. A tan interior debuted on 1994 models. The wide center tunnel forces the driver's legs to angle toward the left and requires some fancy footwork while shifting or braking.

and implied that it was ready to rumble on the road or on the track. The "10" stood for the number of cylinders in the Viper's ample 8.0-liter (488 cubic inches) powerplant. Stretched around this V-10 was a minimalist but muscular-looking body that was as wild looking as any of Chrysler's cars from the 1960s. Chrysler unveiled the Viper concept in January 1989; in May 1990 it was approved for production, and Chrysler was back in the musclecar business.

Chrysler also re-entered the muscletruck business in the 1990s. In 1978 the company was way ahead of the sport-truck curve with the *Li'l Red Express* truck, which was notable thanks to a 360-ci V-8, and two tall, chrome "big-rig" exhaust stacks that sprouted up on each side of the cab. With the new Dodge Ram SS/T, the stacks were gone, but the "big-rig" look was back.

Dodge Viper RT/10

When it gets down to priorities, Mopar muscle machinery has emphasized raw power over style and sophistication. This tradition lives on in the Dodge Viper RT/10. Like a gunslinger ready to draw, the Viper's intimidating posture crouches low to the ground with its huge wheels spread in a wide stance. It peers at you with slanted evil-eye headlights. Muscular, bulging fenders dare you to take a ride. The RT/10 has an outlaw attitude to match its lean and mean looks. Coiled up inside the engine compartment is an 8.0-liter (488 cubic inches) V-10 engine that can unleash sufficient amounts of torque and horsepower to leave musclecars past and present cowering in the shadows.

While most engineers behind today's muscle machines take great pride in their sophisticated approach to high performance, Chrysler designed the RT/10 as a throwback to the wild and woolly Shelby Cobra. The original RT/10 (RT stood for Road and Track, and 10 was the number of cylinders) was unveiled on January 4, 1989, at the Detroit auto show. Its purpose was to be a one-of-a-kind concept car to gauge how willing the driving public was to trade in comfort and convenience for outrageous performance. The public turned out to be very willing. Before the Viper display came down at the end of the show, Chrysler's order books had a list of customers ready to buy an RT/10.

It took the next three years for Team Viper (a handpicked group of about 100 Chrysler engineers, designers, and managers who volunteered) to develop the concept car into a limited-production sports car. Team Viper acted as an independent group within Chrysler to the point of selecting their own suppliers. Official approval for production of the aluminum V-10–powered sports car was announced in May 1990. A year later, the Viper hit the bricks as the official pace car of the Indy 500. The first of 285 1992 Viper RT/10 roadsters rolled off the assembly line in December 1991. The base price was $50,000 plus $700 in destination charges, $2,600 in tax penalties for being a gas guzzler, and $2,280 for being an over-$30,000 luxury car.

Anyone who expected to play with a car delivering 400 horsepower at 4,600 rpm and 465 ft-lb of torque at 3,600 rpm should expect to pay for guzzling fuel, but a luxury tax added insult to injury on a car that had a skimpy folding top with ill-fitting

At home on the road or at the track, the Viper RT/10 Roadster lives up to the Mopar legacy of powerful R/T models from the original musclecar era. While red was the only color originally available, this 1995 roadster is one of 307 that year to sport the Emerald Green exterior that first became available in 1994. Side curtains and a snap-on top are more reminiscent of 1950s' British sports cars than 1960s' American iron.

plastic zippered side curtains. Even MG owners smirked at the Viper's protection from the elements. Red was the only exterior color available. Air conditioning was an option, although a premium AM/FM/cassette stereo sound system was standard. A leather-wrapped steering wheel and shifter knob, plus leather seat surfaces and white-faced instruments, were the most luxurious features on the initial RT/10. The interior accommodations were also on the skimpy side. In order to get into the vehicle, you had to squeeze through the narrow door opening and climb over the wide door sill to fall into the high-back bucket seats. The cramped quarters have you positioned at an angle to the foot pedals. Take a stab at the clutch with your left foot and you are more than likely to hit the brake.

Hitting the brakes—albeit an excellent four-piston front caliper, four-wheel vented disc system—was not a high priority among most of the Viper buyers. Pressing the pedal to the right opened the door to the

The GTS paced the Indy 500 in May 1996, the same month its production began. Based on sneak previews at the Los Angeles auto show and the Viper Owner Invitational, over three-quarters of the 1,700 coupes built during the 1996 calendar year were presold. Note the NACA duct air intake leading directly to the air cleaner. The twin-bubble roof provides additional headroom, and the front fascia has smaller openings and an extended lower lip compared to the roadster.

Viper's magic. Judicious use of your right foot could, if you didn't slew sideways, get you from 0 to 60 miles per hour in 4.6 seconds on your way to a top speed of 165 miles per hour. In musclecar numbers, that comes out to 12.9 seconds at 113.8 miles per hour in the quarter-mile. For comparison, *Motor Trend* tested a 425-horsepower, 427-ci Cobra street version in September 1966, which netted times of 0 to 60 in 5.3 seconds and 13.8 seconds at 106 miles per hour through the quarter-mile. Wheel spin off the line was a problem with the Cobra. It was a notorious problem with the Viper despite gigantic Michelin tires sized at P275/40 ZR17 (front) and P335/35 ZR17 (rear) on six-bolt 17-inch aluminum wheels measuring 10 and 13 inches wide, respectively.

Team Viper continues in its quest to convert Michelin's best efforts into tire smoke. The 2000 Viper RT/10 has 450 horsepower at 5,200 rpm. Torque is up to 480 ft-lb at 3,600 rpm with over 60 percent of that available just barely past idle at 1,200 rpm. According to *Motor Trend*, this was good enough to knock off 0.4 seconds on the climb to 60 miles per hour while motoring down the quarter-mile in 12.3 seconds at 118.2 miles per hour. Don't even think about coming anywhere close to within a second of those times in any of the original musclecars in full street trim.

Besides becoming more powerful, the Viper has undergone some other significant changes. Black became a second exterior color offering for the 1993 model year, followed by bright yellow and emerald green in 1994. In November 1995, production was moved to the Conner Avenue Assembly Plant in Detroit, where 200 workers assemble each car with pride. The big news for 1996 was the debut of the GTS Coupe, but the RT/10 also benefited by receiving an all-aluminum version of its four-wheel independent suspension, and the re-engineering of its tubular space frame with center spine structure. In all, 200 pounds were shaved from the RT/10.

In 1999, wheel diameter was increased to 18 inches with corresponding tire sizes increased to P275/35 ZR18 (front) and P335/30 ZR18 (rear). The RT/10 has also become more luxury oriented.

The interior of a GTS is more "user friendly" than a roadster, with adjustable pedals allowing 4 inches of travel, more than 2 inches of additional seat travel, and more supportive seats with lumbar adjustment and perforated leather seating surfaces. Power windows and electric door locks and latches were also part of the GTS package.

An upscale cognac-colored Connolly leather interior package is an option, along with a removable hardtop. Standard equipment now includes air conditioning, a 200-watt stereo with CD player, and driver-adjustable foot pedals. The base price for a 2000 model is $70,000.

In a little over a decade, that "one-of-a-kind" RT/10 has been joined by about 10,000 Vipers around the world. This is good news for anyone who

2000 VIPER RT/10 AND GTS SPECIFICATIONS

Body/Chassis	Resin transfer-molded (RTM) composite body panels with sheet-molded compound (SMC), forward-tilting hood/fender assembly/Tubular space frame with center spine structure
Engine	8.0-liter (488 ci) OHV V-10, cast-aluminum block with cast-iron cylinder liners, aluminum heads and oil pan, forged-aluminum pistons, forged-steel connecting rods, and 9.6:1 compression ratio
Transmission	Six-speed manual, limited-slip differential, 3.07:1 final drive
Power Ratings	450 horsepower @ 5,200 rpm 490 lb-ft torque @ 3,700 rpm
Brakes	Four-piston front calipers, single-piston rear 13-inch vented rotors front and rear
Wheels/Tires	Front: 10x18-inch; P275/35 ZR18 Rear: 13x18-inch; P335/30 ZR18
Suspension	Four-wheel independent with cast-aluminum control arms; coil-over shocks
Wheelbase	96.2 inches
Length	176.2 inches (RT/10) 176.7 inches (GTS)
Height	48 inches
Curb Weight	3,460 pounds
Weight Distribution, Front/Rear	48/52

All body panels were either new or modified for the GTS, which, thanks to new aluminum suspension components, weighed about 60 pounds less than previous roadster models. A rear cargo capacity of 20 cubic feet is almost 70 per cent greater than that of the roadster. More rearward weight bias also allowed engineers to dial in more understeer when setting suspension for the coupe.

believes that musclecars are not only part of our automotive history, but also a part of the future.

Dodge Viper GTS

It is amazing what a little resin transfer-molded (RTM) composite surgery can do. When Chrysler designers grafted a sleek fastback roof and tail assembly to the open-top Viper RT/10, they dramatically changed the car's appearance from an ornery-looking hulk of a roadster to a superslick GT. The cosmetically altered car appropriately made its first appearance at the Los Angeles auto show in 1993. Just like the RT/10, the next time the public saw the handsome blue metallic coupe with its wide white racing stripe was at Indianapolis in May 1996, where it served as the official pace car of the Indy 500. Production of the GTS Coupe began the same month it joined the RT/10 Roadster in the 1996 Viper model lineup. Unlike the roadster, which only had 285 units produced in its debut year of 1992, there were 1,166 1996 model year GTS Coupes built, according to the Viper Club of America production registry.

Fortunately, at least as Vipers are concerned, beauty is only skin deep. Under its aerodynamically friendly skin, the GTS is every bit as mean and tough as the RT/10. In order to maintain the extraordinary performance level achieved by the RT/10, it was necessary for Team Viper to implement more than just cosmetic surgery to develop a coupe version. Chrysler claims that the changes resulted in a car that was more than 90 percent new. The 1996 GTS Coupe with air conditioning weighed nearly 100 pounds less than the 1994 RT/10 without air conditioning. Many of the changes developed for the GTS Coupe were also later applied to the RT/10, specifically the conversion to an all-aluminum suspension system and a re-engineered frame that was lighter, yet stiffer and stronger.

The GTS Coupe body was designed to improve aerodynamics as well as passenger comfort and convenience. The twin-bubble roof allows for ample headroom and minimizes wind resistance. Headroom was also the determining factor for using a taller windshield than on the roadster, although the angle, width, and curvature remain the same. The flush-fitting

Wearing red war paint, the 1996 GTS V-10 was over 80 pounds lighter than the 1996 RT/10 engine and produced 8 percent more power. Higher compression, new free-flowing breathing aluminum cylinder heads, a thin-wall aluminum block, revised cam timing, and cold-air intake system all combined for 450 horsepower and 490 foot-pounds of torque.

one-piece glass hatch covers a cargo area that is 70 percent larger than the roadster's. The spare remains a compact 16x4-inch tire, but the roomy rear compartment can transport a full-size rear tire for repairs if necessary. The lower lip of the front fascia was extended to balance the cosmetic and aerodynamic effect of the rear spoiler. Louvers atop the front fenders enhance engine cooling, while a polished aluminum "racing" fuel filler cap recessed into the right rear roof panel simply exudes coolness. As you would guess, the coupe is more slippery than the roadster with a drag coefficient of 0.39 versus 0.50.

Changes were made to the doors on the GTS to accommodate power windows and exterior door handles, neither of which were part of the original RT/10. A neat touch is that the doors are opened from the outside by means of pressing a button, which electrically activates the door latch. Inside, the simple door pull of the roadster is retained. Other concessions to civility in the coupe are standard air conditioning, an increase of about 2.3 inches in seat travel, lightweight but more comfortable seats, and an adjustment knob that allows 4 inches of movement for the foot pedal cluster to increase driver comfort.

As part of the GTS-reducing plan, engineers shaved over 80 pounds of weight from the engine and cooling system. The lightweight engine (thanks to new heads and block) of the GTS also had a higher compression and revised camshaft to put out 450 horsepower compared to the 415 ponies available in the 1996 RT/10. A NACA-design intake duct on the leading edge of the hood rams cooler outside air to the engine. A foam gasket seals matching flanges on the air cleaner and the underside of the intake duct. A unique water separator system in the air cleaner cover deflects incoming air into the engine while employing inertia to force any incoming water to accumulate into a trough where one-way valves, activated by the weight of the water, drain to the ground. A rear-outlet exhaust system, first available on the 1996 RT/10, improved power, lowered cockpit noise levels, and added a throatier rumble to the V-10's sibilant exhaust sound. The Manufacturer's Suggested Retail Price (MSRP) of a 1996 GTS was $66,700, including destination charges. Gas-guzzler

The roadster interior for 2000 is downright lavish compared to the early cars. New cognac-colored Connolly leather interior, including steering wheel and shift knob, is an option, but standard equipment now includes air conditioning, adjustable pedals, and a 200-watt stereo with CD player.

and luxury taxes added another $6,300 to the bill. A 1996 RT/10 was $58,500 without taxes.

In 1997, the RT/10 was upgraded to the 450-horsepower engine along with the power windows and the metallic blue with white racing stripe color scheme of the GTS. In exchange, the GTS could be ordered in red. According to Viper Club of America statistics, the GTS outsold the RT/10 1,166 units to 721 in 1996. Coupe sales were even stronger the next year as GTS sales totaled 1,671 versus 117 for the RT/10. A big surge of the early coupe sales came from RT/10 owners eager to own a matched set of Vipers. Through November 2000, Chrysler has produced a total of 7,191 RT/10s since 1992 and 4,484 coupes since 1996. According to Chrysler's 1999 statistics, 96 percent of Viper owners are male, with the average age being 48 years old and an annual household income of $191,000.

***Left:** The 2000 RT/10 is more refined than the early roadsters. This vehicle has benefited from developments through the years that include a stiffer yet lighter chassis, 18-inch wheels, a removable hardtop, and power windows.*

The GTS went racing in the form of the GTS-R that was first introduced in 1995 as a limited-production car made specifically for worldwide GT racing. Indy car chassis builder Reynard Racing Cars partnered with Team Viper to develop this car. Wearing a carbon-fiber bodysuit stretched over a stress-relieved steel space frame welded onto the stock chassis, the GTS-R came equipped with a standard 525-horsepower V-10 to show off American muscle overseas in European road racing. Optional powerplants were rated at 650 and 750 horsepower. In 1997, a Viper GTS-R won the *Federation Internationale de l'Automobile* (FIA, the international motorsport ruling body) GT2-class championship for production-based sports cars. In 1998 the GTS-R won its class at the legendary 24 Hours of Le Mans, a feat repeated in 1999 and 2000. To commemorate its racing success, Dodge built 100 white-with-blue-stripe GT2 Championship Edition Coupes in 1998. Among the modifications to these specials were a GTS-R–style biplane rear wing, a 460-horsepower V-10, and 18-inch BBS wheels. The GT2 listed for $85,200, including shipping.

In 1999, the ACR, basically a GT2 minus the rear wing and special paint scheme, became part of the Viper lineup. The ACR (American Club Racer) is primarily aimed at sports car club racers who need something to race on Sunday and still drive to work on Monday. ACR engines have 460 horsepower and develop 500 ft-lb of torque at 3,700 rpm. At 3,356 pounds, the ACR weighs a little over 100 pounds less than a standard Viper thanks to the deletion of the driving lamps (replaced with air intakes), air conditioning, and the audio system. The latter two items remain as options. The ACR package includes five-point safety belts, K&N filters, aluminum-bodied Koni shocks, one-piece BBS forged-aluminum wheels with "Viper Head" center caps, special badging, and 2.25-inch-internal-diameter racing springs. The ACR sits an inch lower than a stock Viper. A 2000 model ACR lists for an MSRP of $72,225, plus $775 in destination charges.

A comparison of test numbers carried out by *Motor Trend* revealed that Team Viper has done an excellent job of keeping performance levels remarkably close between the roadster and various coupe iterations. The

Despite the luxury touches and toned-down exterior color, Viper RT/10 is still a beast with 450 horsepower on tap to propel it from 0 to 60 miles per hour in 4.2 seconds, with quarter-mile elapsed times in the low 12s.

1999 GTS turns in a 0-to-60-mile-per-hour time of 4.1 seconds with a quarter-mile posting of 12.2 seconds at 119.8 miles per hour. A 2000 RT/10 comes in 0.1 second slower in both tests, while the GT2 model was quickest at an even 4 seconds in the run to 60 miles per hour. At the drag strip the GT2 clocked 12.1 seconds at 120.5 miles per hour.

As good as these numbers are, the great news for musclecar aficionados is that Team Viper is still looking at ways to improve the Viper.

Plymouth Prowler

The Prowler is a factory hot rod that truly looks the part. Following in the wide tire treads of the Viper, it started out as a radical concept car that created such an emotional response among car lovers that Chrysler put it into production with as few changes as possible. Definitely more show than go, the Prowler still earns its place in the modern American muscle group through the sheer boldness and audacity it represents on Chrysler's part to actually sell a car so far out of the mainstream of production vehicles.

The Prowler first appeared in January 1993 as a concept car at the North American International Auto Show in Detroit. The purple metallic nouveau hot rod struck a nerve with the public. Over the next year, Chrysler was besieged with around 130,000 inquiries regarding the availability of the Prowler. The company did not need that much of a shove to get moving. Shortly after the Detroit Auto Show, feasibility studies began to determine how much of the show car could be transferred into a production model. Just as Team Viper recruited employees who were sports car and performance enthusiasts to produce the Viper, Team Prowler put together workers who loved hot rods.

Three years later, in January 1996, the Detroit Auto Show was the site of another Prowler preview. This time, it was a not-so-sneak preview at the Prowler that would be available at Plymouth dealers. Unfortunately, this long-lead preview allowed for a buyers' frenzy and market speculators' price-gouging to ramp up (increase) well ahead of the car's actual production, which did not begin until July 1997. Besides following the path of the Viper in transitioning from concept car to road car, the Prowler is built at the Conner Avenue Assembly Plant in Detroit alongside the Viper production line. Limited production and high demand made the early cars ripe picking for speculators who charged anywhere from 50 to 100 percent above the car's base price of $40,280.

Just as the original hot rod has its roots in Southern California, so does the Prowler. In 1990, a sketch from a brainstorming session at the Chrysler Pacifica West Coast design operation near San Diego caught the eye of Bob Lutz, then Chrysler president. Former Design Vice President Tom Gale, who was building his own hot rod at the time, enthusiastically supported the project of a Chrysler retro hot rod. It was Gale who nixed plans for a retractable hard top, insisting that real hot rods were roadsters. It was also Gale who, in searching for paint for his own rod, came across the metallic hue that evolved into Prowler Purple.

The Prowler's other distinctive features are its cycle fenders, front bumpers, and headlights. The bumpers could have been left off the concept car, but

Is this a face that only a hot rodder could love? Apparently not, as the 1993 debut of the Prowler concept car drew over 130,000 inquiries within a week's time. With the exception of revised headlights and body modifications for side-impact protection, the 1997 street car remains remarkably close to the original concept that was designed with front bumpers for highway use.

the plan was to design a car that could be produced some day. The bumpers consist of hollow aluminum extrusions welded together and are covered with injection-molded urethane. The bumpers are a bit larger and protrude forward on the street version to meet safety standards. Safety and practicality issues led to the redesigned headlight system. Projection lights, which do not require large reflectors, met lighting standards while retaining the streamlined placement on the cowling. The movement of the fenders with the front wheels affected where the side marker lights were positioned for complete visibility.

Other problems encountered in the Prowler's transformation to the real world included widening the body 3 inches to accommodate side-impact protection, and increasing the wheelbase to 113 inches from 111.5 inches to improve the visual impact. The need for a wider radiator affected the length and shape of the hood. The windshield needed to be altered so that windshield wipers could be added. Run-flat tires were specified because there wasn't room to carry a spare tire.

Real hot rods are assembled from a combination of donor parts. Following this tradition, Team Prowler raided the Chrysler parts bins for its factory hot rod. A minivan steering gear was modified for

Originally equipped with a 214-horsepower V-6 and an automatic transmission, attempting a burnout from the Prowler's 20-inch rear tires was futile. For cruising the local drive-in, nothing could touch the Prowler. The demand was so high that early cars commanded 50 to 100 percent over the $40,280 base price.

2001 PLYMOUTH PROWLER SPECIFICATIONS (1997 SPECS IN PARENTHESES)

Body/Chassis	Two-passenger, two-door aluminum and composite
Engine	3.5-liter SOHC 24-valve aluminum V-6, fuel injection with stainless-steel exhaust headers; 10.0:1 compression ratio
Transmission	Four-speed, fully adaptive, electronically controlled automatic with AutoStick manual control; 3.89:1 final ratio
Power Ratings	253 horsepower @ 6,400 rpm (214 horsepower @ 5,850 rpm) 255 lb/ft torque @ 3,950 rpm (221 ft-lb torque @ 3,100 rpm)
Brakes	Power assisted four-wheel vented discs, 11.1-inch, front; 13.0-inch, rear
Wheels/Tires	Front: 17-inch; P225/45 HR17 run-flat Rear: 20-inch; P295/40 VR20 run-flat
Suspension	Front: Upper and lower control arms with pushrod, coil-over shocks, sway bar Rear: Independent multilink, short/long arm, coil-over shocks, sway bar
Wheelbase	113 inches
Length	165 inches
Height	51 inches (top up)
Curb Weight	2,850 pounds

use on the Prowler. Other cannibalized components include Cirrus/Breeze suspension parts for the rear, and Viper front and rear springs and shocks. The 1996 Grand Cherokee was the source for the steering column shroud and steering wheel. The steering wheel received a unique air bag covering and horn pad. Various switches and interior bits were borrowed from the LH, Neon, and Cirrus/Breeze.

The largest and most controversial borrowed component used in the Prowler was the 3.5-liter 24-valve SOHC V-6 engine that came out of the LH Series sedans. Mating this engine to a four-speed automatic AutoStick transmission added insult to the injury of foregoing V-8 power. For some old-school hot rodders, trying to coax the 20-inch-tall rear tires into a smoky burnout with only 214 horsepower was not the way to impress the kids down at the local drive-in. Tom Gale defended the choice and said a V-8 would have intruded too far into the passenger compartment. Correcting this would mean altering the wheelbase beyond what was acceptable. Gale said that a V-6 with a

The biggest engine Chrysler could squeeze under the purple metallic hood of the 1997 Prowler was a 3.5-liter 24-valve SOHC V-6 from the LH sedans. Mated to a four-speed AutoStick transaxle, 0-to-60 times were just over 7 seconds. In 2001, horsepower is up to 253 with a drop in elapsed time to a more respectable 6.3 seconds.

Hot rod roadsters tend to have rather cramped cockpits, and the Prowler is no exception. In true hot rod fashion, Chrysler engineers raided the parts bins for many interior switches and components, including a Grand Cherokee steering wheel and column. A body-colored instrument panel with white-faced gauges and steering column–mounted tachometer are classy touches.

transaxle in the rear fit the Prowler's 1990s attitude of being innovative yet retro. The Prowler's capability of going from 0 to 60 miles per hour in a tick over 7 seconds backed Gale's argument. This was certainly quick enough for what was essentially intended to be a Sunday cruiser and not a dragster.

Everyone applauds the innovative use of materials to manufacture the Prowler. Chrysler claims that almost 900 pounds of the Prowler's 2,850-pound curb weight is aluminum, which makes it North America's most aluminum-intensive vehicle. Aluminum is used for the tub, frame, body sheet metal, and wheels. Other advanced materials consist of a single magnesium casting for the instrument panel; sheet molding compound for the cycle fenders, rear valance, and quarter panels; and polymer for the grille bars.

Pop open the rear-hinged deck lid of a Prowler and anything that is thicker than the Sunday paper will have to go into one of the few options offered. Originally priced at $4,600 ($5,075 in 2001), a "pup" trailer that mimics the Prowler's rear styling down to the taillights and rides on unique 15-inch wheels is perfect for long-distance cruising.

Through November 2000, over 6,500 Prowlers have been produced. Other colors began appearing with the 1999 model year. The most popular color as of September 2000 is black (1,839), followed by red (1,385), purple (1,354), yellow (959), silver

(688), and limited-edition black/red two-tone vehicles (100). For 2001, a black/silver "Black Tie edition" will be offered along with Prowler Orange. Red, yellow, black, and purple exterior colors have been discontinued. The MSRP for the 2001 Prowler is an even $45,000 ($44,225 plus $775 shipping).

The days of Prowler Purple may be gone, but so is the anemic horsepower rating. The latest version of the 3.5-liter V-6 belts out 253 horsepower to get 0-to-60 times down to 6.3 seconds with a quarter-mile turned in 14.9 seconds. The Plymouth division joins Prowler Purple in the history books as well. For 2001, can you say Chrysler Prowler?

Dodge Dakota R/T

Those who scoff at a pickup truck, especially a compact one such as the Dodge Dakota R/T, being considered a performance vehicle need to take a refresher course in musclecar history. Musclecars derived from the compact cars that followed the first wave of import-car hysteria that struck the U.S. auto industry in the late 1950s. The Dodge Dakota R/T debuted in 1986 to fill the niche between full-sized domestic pickup trucks and the minipickups that were an offshoot of the second import assault on Detroit that began in the early 1970s. The Dakota was the first compact pickup. It was easier to drive and park than a full-sized truck, yet it was built like a "real truck" to outhaul or out-tow the smaller trucks.

Competent but conservatively styled, the Dakota received a major restyle in 1997 when it took on the macho "semi tractor" look of its bigger brother, the Dodge Ram. Light-truck sales had taken over half of the new-car market. Some of this market surge came from young people who had grown up driving minitrucks and were looking for something larger, more powerful, and sporty. Something like a musclecar with a cargo bed.

Enter the Dodge Dakota R/T in the spring of 1998. The name (R/T stands for Road and Track) evoked memories of the powerful Dodge cars of the musclecar era such as the Charger R/T and Challenger R/T. The V-8 could send the rear tires squirming sideways with a good shove on the gas.

About 900 pounds of aluminum are included in the Prowler's 2,850 pounds. A special polymer, however, was used to craft the grille. One of the biggest challenges in going from the concept car to a street-legal hot rod involved the headlights. Special projection-beam headlights solved this problem.

The Dakota's 250-horsepower 5.9-liter Magnum V-8 may come up a bit short in both cubic inches and horsepower compared to its ancestors, but a cargo bed full of air does not do much for rear-wheel traction, so the thrill factor is definitely worthy of the R/T model designation.

The V-8 had a performance-tuned dual inlet stainless-steel exhaust system to give it a bit more character than the more mundane versions found in the Durango and Grand Cherokee, and a heavy-duty electronic four-speed automatic transmission (46RE) to handle the 345 foot-pounds of torque. The Dakota R/T also featured 17-inch aluminum wheels, P255/55 R17 tires, a 1-inch-lower sport suspension,

The 1997 Dodge Ram SS/T evolved from the 1996 Indy Ram trucks that were painted to match the blue-and-white Viper GTS Indy 500 pace car. Dodge discovered that it was the 245-horsepower V-8 as much as the paint job that led to the sold-out run of 5,000 1996 trucks.

1997 DODGE RAM SS/T SPECIFICATIONS

Body/Chassis	Steel cab/ladder-type frame
Engine	5.9-liter OHV V-8 with sequential, multiport fuel injection
Power Ratings	245 horsepower @ 4,000 rpm 335 ft-lbs torque @ 3,200 rpm
Transmission	Four-speed heavy-duty automatic
Brakes	Front disc/rear drum with ABS
Wheels/Tires	17-inch aluminum alloy; 275/60 R17
Rear End	Rear drive, limited-slip differential, 3.90
Suspension	Front: Double wishbone independent, coil springs, gas-charged shocks, stabilizer bar Rear: Live axle, longitudinal leaf springs, gas-charged shocks
Wheelbase	118.7 inches
Length	204.1 inches
Curb Weight	4,216 pounds
EPA Fuel Economy, City/Highway	13/17 miles per gallon

front and rear anti-sway bars, and a limited-slip differential. Actually, the engine ($1,585) and transmission ($950) were considered "mandatory options" for the R/T Sport Group package ($2,320), which consisted of the wheels, tires, suspension, an interior/exterior light group, high-back bucket seats, heavy-duty sound insulation, body-color front and rear bumpers, various interior upgrades, and special Dakota R/T badging.

The R/T could only be ordered on two-wheel-drive models in either Regular or Club Cab versions. The colors were limited to Intense Blue, Deep Amethyst, Flame Red, and Black. A base model for $13,260 could not be ordered and have the big engine stuffed inside. Once you added in air conditioning, a killer stereo, power windows, power locks, and other optional necessities, the Dakota R/T sticker would total around $23,000 to $24,000. This was not a bad deal considering what it buys in terms of muscular performance, looks, and utility.

Dodge says about 2,000 Dakota R/Ts were produced in 1998 with 1999 production around 5,000 units. Club Cab models were more popular, especially in 1998 when they outsold Regular Cabs about four to one. Solar Yellow was added to the color choices in 1999. It was also available in limited quantities on early-2000 production models. Bright White and Amber Fire were added in 2000, as Deep Amethyst was deep-sixed. Bright Silver was added as an exterior color for 2001.

The Dakota R/T has built up a popular cult following supported by an aftermarket offering superchargers and other ways of increasing the truck's performance. Right out of the box, it has a 0-to-60-mile-per-hour acceleration of 7 seconds flat with quarter-mile runs of 15.4 seconds at 89.0 miles per hour, according to *Motor Trend*'s testers. The NHRA has started a Pro Stock Truck class where the Dakota is a top runner. The Dakota R/T may be a compact truck, but it can handle the weight of the Mopar performance tradition.

Dodge Ram SS/T

The 1997 Dodge Ram SS/T was a follow-up to the surprising success of the limited-production run of Indy Ram trucks Dodge had offered the preceding year. In 1996, Dodge built 5,000 special-edition, metallic blue and white striped, Ram 1500 pickups that emulated the support trucks that were supplied along with the Viper GTS pace car for the Indy 500. At first, Chrysler officials thought the racy Viper paint scheme or the tie-in to the Indy 500 had created the

The 5.9-liter V-8 in the Ram SS/T produced 245 horsepower and a mellow rumble through the high-flow exhaust system. It was actually an extra $860 option, as was the $960 heavy-duty four-speed automatic transmission and the $1,360 SS/T package that included a heavy-duty suspension and 17-inch wheels to go with the monochromatic paint and Viper stripes.

Based on the upscale Laramie trim package, the SS/T interior included a premium cloth 40/20/40 bench seat with center console, air conditioning, and full power accessories. The exterior colors available were Red, Emerald Green, and Black with silver stripes; or White with blue stripes.

unexpected rush for these trucks, but further analysis showed that the performance aspects of the 245-horsepower 5.9-liter V-8 and the sporty 17-inch aluminum wheels and tires played a big part in the truck's appeal. The 1997 Ram SS/T was essentially the same package with a slightly broader exterior color chart.

Dodge had been down the muscletruck route before. In 1978 the company was way ahead of the sport-truck curve when it offered the *Li'l Red Express* truck. This was a limited run of D150 stepside bed, full-size pickups powered by the venerable 360-ci V-8 with a modified automatic transmission. The carbureted police special engine put out 225 horsepower. *Hot Rod* magazine tested one of these trucks in November 1977 and recorded a quarter-mile run of 14.7 seconds at 93 miles per hour. It was one of the fastest domestic vehicles for sale thanks to the lax emissions requirements of the period regarding trucks. The bright red trucks had no catalytic converters or smog pumps. Two tall, chrome "big-rig" exhaust stacks sprouted up on each side of the rear of the cab. Varnished wood trim gave the bed and tailgate a "woody" look. Dodge sold a little over 7,000 of these beauties in 1978 and 1979.

The big-rig look of the 1970s was still a hit with pickup buyers as evidenced by their positive response to the brawny fenders and hulking grille of the Ram models that first appeared in 1994. The 1997 SS/T captured high-performance visual appeal with a body-color-matched grille and bumpers, Viper-style racing stripes, fog lights, chrome exhaust tip, and 17-inch aluminum wheels. The exterior colors consisted of Black, Emerald Green, and Flame Red, all with wide silver stripes, or Bright White with blue stripes. Inside, a tachometer and a premium cloth 40/20/40 bench seat with a wide center compartment/armrest large enough to stow a laptop computer were also included.

The heart of the SS/T was under the hood. The engine measured 360 cubic inches (5.9 liters). A specially tuned version of the electronic fuel-injected 5.9-liter V-8 developed 245 horsepower as it let out a mellow roar while breathing through a high-flow exhaust system. A four-speed, heavy-duty automatic transmission delivered the 245 horsepower and 335

The Dakota R/T is notable among high-performance pickups by being available in Club Cab format as well as the Regular Cab. The color choice was also unique when the package was first offered in 1998 with Intense Blue and Deep Amethyst (as shown) joining the more traditional red and black. A lowered suspension (by 1 inch), 17-inch wheels, body-color grille and bumpers, and upgraded interior are also included.

ft-lb of torque through a limited-slip differential with a 3.90 rear end and 275/60 R17 tires. Despite their wide tread and modern construction, the tires still let out a scream when you shoved on the gas.

Motor Trend clocked a 0-to-60-mile-per-hour time of 7.9 seconds in a 1997 Ram SS/T. At the drag strip, the truck checked in with a 15.8-second run at 84.6 miles per hour. For anyone not impressed with those numbers, an option was the dealer-installed Mopar Performance Magnum R/T performance package, which consisted of matched engine computer, camshaft, intake, and exhaust headers said to deliver another 46 horses.

Although stiffer shocks and springs were part of the SS/T package, the suspension was not lowered from the stock ride height. The SS/T may have been faster than a standard Ram pickup, but its handling was not much improved beyond the extra capabilities provided by wide wheels and performance tires.

The SS/T package cost $1,360 and required the additional special 5.9-liter V-8 ($860) and heavy-duty automatic transmission ($960) at an extra cost. The Ram 1500 Laramie SLT to which all this was added had an MSRP of $20,025 including destination fees. The Laramie was a well-equipped vehicle, and air conditioning, tilt wheel, cruise, and full

The 5.9-liter V-8 and heavy-duty four-speed automatic transmission were considered mandatory—but extra cost—options on the Dakota R/T. With 250 horsepower on tap and a limited-slip differential, the truck was good for 7-second runs from 0 to 60 miles per hour.

power accessories were included so a nicely equipped SS/T could be purchased for under $25,000.

The SS/T was an option package instead of a limited-production run. It disappeared from the Dodge catalog in 1999. A similar truck could be ordered, but it took a careful study of the options list. A key ingredient, besides the obvious engine choice, was the Sport Appearance Group (which included 16-inch wheels) that was offered only with Flame Red, Black, Bright White, or Intense Blue exterior colors. Hopefully, muscletruck fans will not have to wait another 20 years for Dodge to follow up on the Ram SS/T.

A choice of cloth bucket seats or a bench seat, tilt wheel, cruise control, and deluxe lighting are part of the R/T package.

DODGE DAKOTA R/T SPECIFICATIONS – REGULAR CAB (CLUB CAB)

Body/Chassis	All-steel body on ladder-type steel frame
Engine	5.9-liter (360 ci) OHV, cast-iron-block V-8 with sequential, multiport electronic fuel injection; 8.9:1 compression ratio; regular unleaded fuel
Power Ratings	250 horsepower @ 4,400 rpm 345 ft-lb torque @ 3,200 rpm
Transmission	Automatic, four-speed overdrive
Brakes	Front disc/rear drum with ABS
Wheels/Tires	17x9-inch cast-aluminum; P255/55 R17
Rear End	Limited-slip differential; 3.92 final drive ratio; two-wheel drive only
Suspension	Front: Double wishbone independent, coil springs, gas-charged shocks, 15-millimeter stabilizer bar Rear: Live axle, four-leaf longitudinal springs, gas-charged shocks, 19-millimeter stabilizer bar
Wheelbase	112 inches (131 inches)
Length	196 inches (215.1 inches)
Curb Weight	3,924 pounds (4,110 pounds)
Bed Length/Payload Rating	6.5 feet/1,275 pounds
EPA Fuel Economy, City/Highway	12/16 miles per gallon

CORVETTE

3

THE GENERAL'S SPECIAL FORCES FOR THE 1990S

In the 1960s, General Motors offered the widest selection of musclecars, with every division except Cadillac producing at least one high-performance model. By the late 1970s, thanks to emissions and fuel economy concerns, most of these models had disappeared. Those that had soldiered on, including the Corvette, suffered from severe muscular atrophy.

By 1990, the GM musclecar ranks had dwindled to the Corvette, Camaro, and its Pontiac sibling, the Firebird. The bright side was that computer technology, which enabled emissions and fuel economy friendly technology to coexist with power enhancements, had led to a resurgence of performance options. The Corvette led the way with development of the ZR-1, which was designed to establish the Corvette as a worthy equal to the world's highest-performing exotic sports cars.

The Corvette's 40th anniversary was in 1993, and the top of the line that year was a ZR-1 wearing the Ruby Red metallic 40th Anniversary paint scheme. Only 245 cars were sold with this option. The 1993 ZR-1 option package was only available on coupes and cost an additional $31,683.

When the ZR-1 debuted in 1990, it featured unique doors and extrawide (3 extra inches) rear bodywork to clear its 17x11-inch-wide rear wheels. The big wheels were needed to carry the massive Goodyear P315/35 Z-rated tires for what Chevy had planned to be the world's fastest production car.

While pushing forward with advanced technology, GM would also look to its past for inspiration in reviving limited-edition track-ready Corvette models such as the Grand Sport and Z06. On the Camaro side of the company, the Z28 name would also resume its lofty performance position.

Another high-performance name of the past was revived with a 1990s attitude in the full-sized Impala SS. The Ram Air option made famous by the GTO was reintroduced with a vengeance as the Firebird Trans Am spread wings that had previously been clipped by emissions regulations.

Performance cars were not the only vehicles to capture GM's attention at the beginning of the 1990s. The light-truck market was gaining momentum, and Chevrolet was gearing up for battle with Ford. If musclecars were useful sales tools during the 1960s, then high-performance versions of trucks could do the same in the 1990s.

Some of the GM vehicles and names have disappeared or have had new faces since the 1960s. Heading into the 1990s, GM was ready to start dealing out high-performance musclecars again.

Corvette ZR-1

It's good to be king. For Chevrolet performance addicts, the Corvette reigns over all. That's why the Corvette special edition built in 1990 that took on the world's performance leaders was affectionately dubbed "King of the Hill." Corvette Chief Engineer Dave McLellan proclaimed that the "ZR-1 is a Corvette, only more so."

The ZR-1 Corvette option offered from 1990 to 1995 appeared in the Chevy catalog under the listing

The ace in the hole for the ZR-1 "King of the Hill" Corvette was an aluminum, DOHC, 32-valve LT5 V-8 developed in conjunction with Lotus Engineering. Initially developing 375 horsepower, the sophisticated powerplant had exceeded Chevrolet's goal of 400 horsepower with a rating of 405 horsepower by 1993.

Special Performance Package. It was a rather understated description for a $27,016 (1990) to $31,258 (1994–1995) bundle of goodies that not only consisted of major mechanical changes to the regular-production Corvette, but a major philosophical shift in how those changes were developed. Adding to the heft of the ZR-1 package were the world-class performance hopes and aspirations of Corvette fans inside and outside of GM.

The ZR-1 was intended to be the world's fastest production car. This vehicle was America's answer to high-performance and higher-priced foreign competitors such as Porsche, Ferrari, and Lamborghini. The ZR-1 was not only significantly cheaper than its rivals, but it also easily met all existing emissions, noise, and safety laws and retained day-to-day drivability. It also burned regular unleaded fuel to avoid gas-guzzler taxes.

At the heart of this symbol of American high performance was a small-block 5.7-liter V-8. The red-white-and-blue DNA of the ZR-1's engine was not in the form of stars and stripes, but the crossed bars of the Union Jack. The LT5 of the ZR-1 had an aluminum block and heads, four overhead camshafts, 32 valves, and was designed by Lotus in England. The engines, however, were made in the United States by Mercury Marine in Stillwater, Oklahoma. GM's Lotus engineering group was chosen because of their experience with high performance and racing four-valve overhead cam head engines. Chevy U.S. engineers consulted with Lotus during development. As for the manufacturing site, Mercury was chosen based on the company's modern manufacturing facilities for building aluminum marine engines, and because only 18 LT5s would be turned out in a day versus the 1,000 or so a GM assembly plant was geared to do.

During the 1980s, Chevy had experimented internally with turbocharging in its quest for the "ultimate" Corvette. Fuel consumption, reliability, or emissions were usually not up to the standards set for a car that still had to be a capable daily driver. They even went outside GM in offering a 345-horsepower twin-turbo package, installed by Callaway Engineering, as an option from 1987 to 1991. Eventually, computer technology such as direct-fire, distributorless ignition systems, electronic spark control with knock sensors, and electronic three-phase, twin-injector sequential fuel injection could carry out what Chevy engineers called a "bi-modal" or dual-personality vehicle. This

The Grand Sport replaced the ZR-1 in 1996, and it offered similar performance at about half the price. With a name and paint scheme that paid homage to special lightweight racing versions of the 1963 Sting Ray, the Admiral Blue with white stripe Grand Sport option ($3,250 for coupes), unlike the ZR-1, was available as either a coupe or convertible. Black-painted 17-inch wheels wore the same-size rear tires as the ZR-1.

vehicle could be a refined road car as well as a ferocious racecar based on driver input.

The LT5's induction system consisted of a three-valve throttle body with a small primary valve for quick response and efficient low-speed running with two larger secondaries for full-power operation. The throttle valves work in conjunction with a 16-runner tuned-length intake manifold and a set of two electronically controlled fuel injectors per cylinder. Based on various sensors feeding into the engine's electronic brain, at 3,500 rpm (half throttle) the secondary ports become wide open. In addition, the cams associated with the larger secondary throttle ports have greater timing than those of the primary ports. This provides variable valve timing to optimize airflow and fuel delivery. A secondary benefit of this system is that it allows a power switch, located on the car's center console to "lock out" the secondary port throttle valves, to limit operation to the primary port system. This is a handy feature that denies access to all of the King's horses to parking-lot attendants and other less qualified drivers. Other features of the LT5 include fast burn, center plug, cloverleaf-shaped combustion chambers, and advanced oiling and cooling systems.

Although Corvette engineers were shooting at a 400-horsepower target for the LT5, no one registered disappointment that the LT5 of 1990 to 1992 produced "only" 375 horsepower at 6,000 rpm and 370 foot-pounds of torque at 4,800 rpm. What proved disappointing were the valvetrain problems that delayed the ZR-1 from debuting as a 1989 model. ZR-1 production started in August 1989 with 84 cars built in 1989, but they were classified as 1990 models.

Call them whatever year you like, but call them fast. According to Chevy, the ZR-1 sprinted to 60 miles per hour in 4.9 seconds and hustled through the quarter-mile in 13.4 seconds. Top speed was around 180 miles per hour. *Motor Trend* tested the ZR-1 in April 1990 and achieved a 0-to-60 time of 4.4 seconds. Their quarter-mile test figures got down to 12.8 seconds. For comparison, a *Motor Trend* test

The LT4 5.7-liter V-8 of the Grand Sport may have lacked some of the sophistication of the LT5, but with new aluminum heads, Crane roller rocker arms, and a 10.8:1 compression ratio, it put out 330 horsepower at 6,000 rpm. The Grand Sport was capable of going 0 to 60 in 4.7 seconds, a few ticks slower than the ZR-1.

of the previous "ultimate" Corvette—a 1967 427, 435-horsepower Sting Ray—turned up 0 to 60 in 5.5 seconds and the quarter-mile in 13.8 seconds.

Unlike musclecars of the past, the ZR-1 was more than just a "big-engine" option package. In addition to a six-speed manual transmission, leather sport seats with power adjustment, and a Delco-Bose stereo system, the ZR-1 included the optional (for other Corvettes) FX3 Selective Ride Control that consisted of three modes (Performance, Sport, and Touring) controlled by a console-mounted switch that varied shock dampening rates depending on vehicle speed. Another Corvette option that was standard to the ZR-1 was the Z51 suspension with firmer springs and anti-roll

1990 CORVETTE ZR-1 SPECIFICATIONS

Body/Chassis	Two-door coupe body on welded-steel uniframe
Engine	5.7-liter aluminum DOHC, 32-valve V-8 with sequential fuel injection; 11.0:1 compression ratio
Power Ratings	375 horsepower @ 6,000 rpm 370 ft-lbs torque @ 4,800 rpm
Transmission	Six-speed manual
Suspension	Front: Independent aluminum short/long arm, transverse monoleaf spring, stabilizer bar, electronic [See query in text re: dampers vs. dampeners]dampers Rear: Five-link independent, transverse monoleaf spring, stabilizer bar, electronic dampers
Wheels/Tires	Front: 17x9.5-inch aluminum alloy; P275/40 ZR17 Rear: 17x11-inch aluminum alloy; P315/35 ZR17
Brakes	Four-wheel discs, front: 13x1.1-inch rotors, rear: 12x1.1-inch rotors
Curb Weight	3,465 pounds
Wheelbase	96.2 inches
Length	177.4 inches

The total Grand Sport production was 1,000 cars—810 coupes and 190 convertibles. This vehicle was only available in the 1996 model year, the last year of the C4 platform.

bars. The ZR-1's biggest handling advantage came from its huge rear tires that were the widest-ever Goodyear Eagle unidirectional radials. They measured a hearty P315/35 ZR17 and were speed-rated to 193 miles per hour. The front tires were P275/40 ZR17. A low-tire-pressure monitoring system came standard on the ZR-1.

The wide rear tires affected the ZR-1's appearance as much as its handling. The rear bodywork of the 1990 ZR-1 was made about 3 inches wider to cover the big radials. The ZR-1 tail also had four rectangular taillights. In 1991, this tail treatment became standard on all Corvettes, although the ZR-1 had wider door and rear body panels to cover its 11-inch-wide rear wheels and huge tires. In 1993, improvements were made to the LT5 valvetrain and cylinder head that aided reliability and increased horsepower beyond the magic 400 mark to 405 horsepower, but it was too late to save the ZR-1.

The last ZR-1 rolled off the assembly line in April 1995. Chevrolet produced 6,939 ZR-1s, but 3,049 were built in 1990 and 2,044 the following year. From 1993 to 1995, each year's production was only 448 units. Chevrolet celebrated the Corvette's 40th anniversary in 1993 by offering an optional Ruby Red color scheme with anniversary logos on all models. There were 245 ZR-1s painted in this color scheme.

A lack of distinctive styling, a high price (almost double that of a standard Corvette coupe), and early engine problems were all factors in the King's demise, but the biggest factor was just how good the basic Corvette had become. Despite the aluminum block and heads, the complex valvetrain and other components made the LT5 engine 40 pounds heavier than an LT1 small-block. In 1996, an aluminum-head LT4 version of the LT1 with 330 horsepower became available. Combined with the optional Grand Sport suspension and trim package, it offered similar performance to the ZR-1 for about $25,000 less. The King was dead, but anyone who owned or drove one will chant, "Long live the King."

Corvette Grand Sport

Chevrolet focused on the future when the 1990 ZR-1 "King of the Hill" was designed, but its successor to the Corvette performance throne had an eye on the past. For 1996, Chevrolet introduced the Z16 Grand Sport option package that included a brand-new 330-horsepower version of the venerable small-block V-8.

The Grand Sport name may only be familiar to hard-core Corvette enthusiasts. It was applied to what was to be a limited run of 100 lightweight 1963 Sting Rays built specifically to compete against the Shelby Cobra in SCCA and European road-racing events. Production Corvettes were too heavy to be competitive with the Ford racers, so Zora Arkus-Duntov came up with a competition car that resembled a Sting Ray, but weighed about 1,900 pounds. The first three were powered by an aluminum alloy version of the small-block V-8 punched out to 377 cubic inches (6.2 liters) that generated 480 horsepower. The cars, wearing the blue-and-white livery of the Mecom-Chevrolet racing team, made an impressive debut at the Nassau, Bahamas, Speed Week in 1963. Driven by Roger Penske and Augie Pabst, two of the Grand Sports came in third and fourth in the Governor's Cup race. The only cars to beat them to the finish line were from the prototype class. In 1964, Penske returned to the Bahamas to win the Nassau Trophy race. Unfortunately, only two more cars were built before the GM management ban on racing cut short the Grand Sport's racing career. One of the final two cars has had Chevy race fans speculating what might have been as it was powered by a 427-ci engine.

The modern Grand Sport pays homage to its forefathers with an Admiral Blue paint job bearing Arctic White stripes. The Mecom team used twin red hash marks on the left front fender to differentiate the two otherwise-identical racecars from each other.

Like its racing namesake, the 1996 Grand Sport also relies on a powerful small-block that uses aluminum in its makeup. The LT4 had special high-compression (10.8:1) aluminum heads. The base 1996 LT1 had a compression ratio of 10.4:1 and made 300 horsepower. In addition to the higher-compression heads, the LT4 utilized Crane roller rockers, a recontoured camshaft, a performance crankshaft, and new fuel injectors to make 330 horsepower at 5,800 rpm. The redline was 6,300 rpm as opposed to the LT1's 5,700-rpm redline, so an 8,000 rpm tachometer replaced the standard

Corvettes with the Grand Sport option had a unique serial number sequence. Available interiors were either all black or a red-and-black combination. Perforated leather sport seats carried "Grand Sport" embroidering.

6,000-rpm gauge. According to Chevy, this was good enough to go 0 to 60 in 4.7 seconds and travel a quarter-mile in 13.3 seconds.

Other pieces of the Grand Sport package, unlike the ZR-1, were available on convertibles as well as coupes, and included the five-spoke, 17-inch aluminum wheels (previously used on the ZR-1), which were painted black and had black brake calipers with silver "Corvette" lettering. Coupes received the ZR-1 front and rear tire combination of P275/40 ZR17, P315/35. ZR17 Goodyears but had rear fender flares in lieu of the wider rear panels used on the ZR-1. Convertibles eschewed the fender flares and rode on tires that measured P255/45 ZR17 up front and P285/40 ZR17 in the rear. Coupes also received 11-inch-wide rear wheels.

Perforated-surface sport seats with power adjustment and "Grand Sport" lettering were also part of

The C5 (fifth-generation) Corvette platform debuted in 1997. In 2001, Chevrolet introduced the Z06, which was aimed at weekend racers. It has so many unique features, such as its own suspension setup, insiders refer to it as the C5.5. Available only as a hardtop, the lightweight Z06 weighs 117 pounds less than a stock coupe or convertible. Using a titanium exhaust system saves over 17 of those pounds. The McLaren F1 is the only other production car with this feature.

the Grand Sport option. Buyers could choose from either black or a red-and-black combination. Corvettes with the Grand Sport option had a separate serial number sequence, 1G1YY2251T5600001 through 5601000. Of the 1,000 cars built, 810 were coupes and 190 were convertibles. The Grand Sport option (RPO Z16) added $3,250 to the $37,225 base price of a coupe, and added $2,880 to the $45,060 price tag of a convertible. Because of the added power of the LT4 engine, all Grand Sports were equipped with a six-speed manual transmission.

Grand Sport handling was enhanced by the F45 Selective Real Time Damping package that replaced the FX3 Selective Ride control of the ZR-1. This new system relied on advanced wheel travel sensors that respond to changes in the road every 10 to 15 milliseconds and adjust each shock individually.

The Grand Sport emulated its historic counterpart in terms of lifespan as well. Chevrolet was cooking up an all-new fifth-generation Corvette for 1997 and the Grand Sport option was not on the C5 menu.

(The 1996 Corvette Grand Sport Specifications, except for the engine, were the same as those presented for the ZR-1.)

Corvette C5

The first Corvette was built on June 30, 1953. The first-year production run was 300 cars in 1953. By the end of the 1996 model year, Chevrolet had built 1,088,063 Corvettes. The one-millionth Corvette rolled off the line on July 2, 1992. Despite the relatively large production totals and a history that spanned five decades, until the 1997 model year, there were only four basic generations of Corvettes.

The Z06 has thinner window glass and lighter wheels and tires than other Corvettes. The external distinguishing characteristics include functioning brake ducts in the bodywork ahead of the rear wheels and red brake calipers. All 2001 hardtops are Z06s.

1997 CORVETTE SPECIFICATIONS

Body/Chassis	Fiberglass two-door coupe/unitized steel
Drivetrain	Front-engine, rear-mounted transmission, rear-drive
Engine	5.7-liter aluminum block and head OHV V-8 with sequential fuel injection
Power Ratings	345 horsepower @ 5,600 rpm 350 ft-lbs torque @ 4,400 rpm
Transmission	Rear-mounted; Standard: four-speed automatic with overdrive; Optional ($815) six-speed manual
Suspension	Front: Short/long arm double wishbone, transverse leaf springs, stabilizer bar, height adjustable Rear: Short/long arm double wishbone, transverse leaf springs, stabilizer bar, height adjustable
Brakes	Four-wheel disc with ABS
Wheels/Tires	Front: 17x8.5-inch cast-aluminum; P245/45 ZR17 run-flat Rear: 18x9.5-inch cast-aluminum; P275/40 ZR18 run-flat
Wheelbase	104.5 inches
Length	179.7 inches
EPA Fuel Economy, City/Highway	18/28 miles per gallon (manual); 17/25 miles per gallon (automatic)
Top Speed	172 miles per hour (per manufacturer)

Standard Z06 colors are limited to white, silver, black, and red. Millennium Yellow costs an extra $600, not including any additional fines for speeding tickets it may attract.

The last major platform change was in 1984. When the all-new, fifth-generation (dubbed C5) Corvette appeared in 1997, it was big news for Chevy and car enthusiasts around the world.

Over its long history, Corvette has become the icon of the American high-performance sports car. This placed a dual burden on the shoulders of the men in charge of developing the new C5—Corvette Vehicle Line Executive and Chief Engineer Dave Hill, and Corvette Chief Designer John Cafaro. They had to come up with a car that would satisfy existing Corvette owners and fans weaned on key elements of the car's heritage including the fiberglass body, V-8 power, and a performance/cost ratio well below foreign competitors. They also had to move forward to incorporate state-of-the-art technology and "build quality" to attract new customers.

A good example is the debate that has raged since the 1960s as to whether the Corvette should emulate most of its competitors by placing the engine amidship. Chevrolet has stirred things up through the years by building mid-engine concept cars (the Astro II, 4-Rotor, Corvette Indy) that appear in the motoring press under *National Enquirer*–like banner headlines asking, "Is this the next-generation Corvette?" Dave Hill faced the dilemma of whether to move the engine amidship by employing what GM calls "Voice of the Customer" research. This survey made it clear that Corvette customers strongly believed that a Corvette must have a front-engine, rear-drive power layout. Hill's team built a CERV (Corvette Engineering Research Vehicle) that kept the engine up front, but placed the transmission at the rear, and connected the two via a torque tube. This not only achieved a balanced 50/50 weight distribution, but it freed up additional interior space and utilized the drivetrain tunnel as a reinforcing backbone.

Inside, a choice of all-black or red-and-black trim is available on the Z06. Heavily bolstered leather seats feature Z06 logos on the headrests. The instrument panel sports italicized numerals on a checkered background with a 6,500-rpm redline tachometer.

The stronger frame design made use of the latest in GM manufacturing techniques, and the C5 required 34 percent fewer parts (1,462 fewer than a C4) to make it more rigid and lighter than its predecessors. The full-length perimeter frame uses side rails made of seamless tubular steel. These rails are hydroformed (pressed into shape by a high-pressure hydraulic press developed by GM) as opposed to the former method of welding 14 pieces together to create each side rail.

The stiffer chassis improved the ride quality of the C5 without compromising handling. No parts were carried over to the C5's suspension, which is height-adjustable and fully independent front and rear. At the rear, upper control arms were added to free the half-shafts from the burden of being suspension components.

Compared to the C4, the C5 wheelbase was stretched 8.3 inches to improve ride and stability. A 4.3-inch-wider front track and a 3-inch-wider rear track (equal to the ZR-1) were added to further aid stability. The rear cargo space nearly doubled. Larger door openings, more seat travel, and a 3-inch-wider footwell made the C5 roomier for people as well.

Despite its larger proportions, the 1997 C5 weighed about 69 pounds less than a 1996 C4.

The C5 sticks to its heritage with a small-block V-8. The 5.7-liter LS1, however, writes it own page in the Corvette history book. With 345 horsepower and 350 foot-pounds of torque, it is more powerful than the LT4 of 1996, yet has better fuel mileage. It features an aluminum block and achieves a similar displacement to its predecessors, but gets there by shrinking the bore and increasing the stroke of its cylinders. This provides more room between the cylinders for extra cooling and added strength. The new block is also more rigid due to deep-skirt construction that extends the bottom of the block past the bearing caps. A rigid new oil pan adds more strength. Extended sumps on each side of the pan increase capacity and ensure that the engine is not starved for oil during sharp cornering maneuvers.

A simplified valvetrain puts valves, rocker arms, and pushrods in line to reduce stress and friction while saving weight. Gone are the "Siamese" ports of the LT1 and LT4 engines, where each port has its own air distribution characteristics. The LS1 ports are identical and enhance engine breathing by directing the air in a straighter path to the combustion chamber. A new composite intake manifold, which weighs less than aluminum, is less restrictive and delivers cooler air to create more power. An individual

2001 CORVETTE Z06 SPECIFICATIONS

Body Style	Two-door fixed-roof coupe
Engine	5.7-liter aluminum block and head OHV V-8 with sequential fuel injection, 10.5:1 compression ratio, titanium exhaust system
Power Ratings	385 horsepower @ 6,000 rpm (6,500-rpm redline) 385 ft-lbs torque @ 4,800 rpm
Transmission	Six-speed manual, rear-mounted
Suspension	Front and rear short/long arm double wishbone, transverse leaf spring, stabilizer bar; height adjustable
Brakes	Four-wheel disc with ABS
Wheels/Tires	Front: 17x9.5-inch; P265/40 ZR17 Rear: 18x10.5-inch; P295/35 ZR18
Wheelbase	104.5 inches
Length	179.7 inches
Curb Weight	3,130 pounds
Weight Distribution, Front/Rear	53/47 (stock C5 51/49)

ignition coil at each cylinder provides a hotter spark to raise performance and lower emissions. The LS1 is the first GM engine to feature Electronic Throttle Control (ETC). Drive-by-wire technology eliminates mechanical linkage and allows for better integration with the traction control system in order to reduce wheel slip.

The longer, lower, wider exterior of the C5 cuts through the wind with a 0.29 drag coefficient, which is lower than most Winston Cup cars. The C5 pays its respects to past Corvettes with hideaway headlights, four taillights, and sculpted indentations along its flanks. Inside, the Corvette's musclecar roots are recalled with a twin-pod dash, analog gauges, and a dash-mounted ignition switch.

Despite its increased size inside and out, the C5 is no heavyweight except when it comes to performance. That is the measurement by which a Corvette model is ranked in the history books. The 1984 C4 (first of its generation) weighed 3,200 pounds and had 205 horsepower for a weight-to-power ratio of 15.61:1. The C5 weighs 3,229 pounds and has 345 horsepower for a ratio of 9.36:1. Taking this to the street, or the track as *Motor Trend* did, the C5 can scoot to 60 miles per hour from a standing start in 4.7 seconds. The quarter-mile is run in 13.3 seconds. That's ZR-1 territory.

Corvette Z06

Forget about the saying that those who fail to learn from history are doomed to repeat it. Repeating history is a thrilling exercise behind the wheel of the 2001 Corvette Z06—a car that has learned its lesson from two of the hottest pages in the Chevrolet history book.

Chevrolet started gathering momentum toward the new millennium with an all-new fifth-generation Corvette coupe in 1997, followed by a C5-edition convertible in 1998, and a lower-priced hardtop model in 1999. In 2000, the factory returned to racing with a team of C5-R racecars to challenge the world's best production-based sports cars at Le Mans.

For 2001, everyone gets to go racing with the new Z06. The original Z06 was the brainchild of Zora Arkus-Duntov. The idea was to take the new 1963 Sting Ray and make it ready for racing right off the showroom floor. Starting out with a split-window coupe, Arkus-Duntov applied his magic touch by adding such items as Al-Fin power drum brakes with sintered metallic linings (four-wheel discs were not offered until 1965), heavy-duty suspension bits,

Positraction, and fuel injection. The Z06 package and required options added $2,480 to the coupe's $4,257 base price. Pricey, but what you received was a Corvette that you could literally drive off the showroom floor to the races. That's exactly what happened in October 1962 when a team of race drivers drove their new Z06s from the St. Louis factory to participate in a race at Riverside Raceway in Southern California. In what turned out to be the racing debut of both the Z06 and the Shelby Cobra, one of the Z06s, owned by Mickey Thompson and driven by Doug Hooper, was the winner.

The 2001 Z06 is a car that can be competitively driven by weekend warriors at autocrosses and track events, yet still get to work on Monday. There are so many unique modifications that Corvette Chief Engineer Dave Hill refers to the Z06 as a C5.5 model. Starting with the C5 hardtop, the lightest Corvette available, Hill's people set up a weight-reduction program that included a titanium exhaust system, lighter wheel/tire sets, and thinner glass for the windshield and rear backlight. The Z06 is 36 pounds lighter than a normal hardtop and weighs 117 pounds less than a coupe or convertible. The titanium exhaust system saved 17.6 of those pounds. The Z06 is the first mass-production car to have such a system. The only other production car to feature a titanium exhaust system was the million-dollar McLaren F1 super exotic car.

All 2001 Corvettes, including the Z06, received a second-generation Active Handling system, which was less intrusive and better integrated with electronic traction control. The key points of the new system were sideslip angle rate control, which responds according to the driver's reaction to sudden handling maneuvers, and the ability to estimate the coefficient of friction on the road's surface. An important feature for Z06 drivers out on the racetrack is the Competitive Mode button on the console, which, when held for 5 seconds, switches off the traction control portion of the Active Handling system to allow some wheel spin.

The Z06 has its own suspension setup that is not available on other models. The FE4 suspension has a larger, hollow front stabilizer bar; stiffer rear leaf spring; and more aggressive front and rear camber settings. The latter aids tire grip during cornering.

The tires of the Z06 are also unique. Developed specifically for the car by Goodyear, the Eagle F1 SC tires have asymmetrical treads and are much lighter than the run-flat tires that are standard on the other C5s. The result is a reduction of 23.4 pounds of unsprung weight to improve handling response. The C5 Corvette already saves weight by not carrying a spare tire, and compensates for this with run-flat

Underneath its red engine covers, new aluminum alloy pistons, revised combustion chambers, a high-lift cam, stronger valve springs, and larger injectors are pumping and squishing away to generate 385 horsepower at 6,000 rpm in the latest LS6 5.7-liter V-8 powering the Z06. Coupled to a new M12 six-speed manual transmission, the LS6 gets the Z06 from 0 to 60 miles per hour in 4 seconds!

tires. The Z06 tires are not of run-flat construction, so a tire-inflating device filled with a special latex sealing compound is part of the Z06 equipment list. The tires ride on special lightweight five-spoke wheels with a light gray metallic finish that are 1 inch wider than stock.

The Z06's engine bears a name out of the Chevy hall of fame. The LS6 designation was originally assigned to a 1971 454-ci big-block that produced 425 (gross) horsepower. The new LS6 is based on the aluminum-block LS1, but it is a screamer. New aluminum alloy pistons, revised combustion chambers, a high-lift cam, stronger valve springs, and larger injectors are pumping and squishing away under its red engine covers to generate 385 horsepower at 6,000 rpm, and an equal amount of torque at 4,800 rpm. A new transmission, the M12 six-speed manual, is geared to assist the high-winding LS6 to take the Z06 from 0 to 60 miles per hour in 4 seconds flat. Chevy claims its top speed is a bit over 171 miles per hour. Their quarter-mile figures are 12.6 seconds at 114 miles per hour. Compare this to the 1971 LS6 (big-block) Corvette, which went 0 to 60 in 5.3 seconds and turned the quarter-mile in 13.8 seconds at 105 miles per hour. A 2001 LS1–powered C5 does the quarter-mile in 13.22 at 109 miles per hour. Road racers should note that the Z06 can corner at over 1 g. The base price for a Z06 is $47,700. At a price that would only be a down payment on cars it can easily outrun, Chevy should have no trouble selling many more than the 199 1963 Z06s it sold.

Chevrolet Impala SS

Contrary to the current practice of dusting off high-flying model names including Malibu, Monte Carlo, and Impala to add luster to low-flying

The SS initials returned to the Impala lineup for the first time in 25 years with the limited-edition 1994 Impala SS model that was first displayed at the SEMA show in 1992. An all-black monochromatic paint job and 17-inch wheels removed much of the dowdiness from its Caprice Classic foundation.

A rear spoiler and retro Impala SS logos were the only ornamentation on the Impala SS. Black was the only color available in 1994. For 1995 and 1996, black cherry and dark gray-green metallic were added. A shortage of wheels limited first-year sales to 6,303 cars. Chevy sold almost 70,000 Impalas before pulling the plug in 1996.

family-rated machinery, Chevrolet in 1994 soared into the high-performance market with an R-rated special-edition Impala SS. Based on the police package version of the rear-drive Caprice Classic full-size sedan, and with a 260-horsepower Corvette-derived V-8, the Impala SS rekindled warm memories of the 1960s—an era when the chrome antelope leaping over twin *S* initials was a symbol for super-sized domestic high performance.

The Impala name goes way back to 1958 when it debuted as a high-end trim option on Bel Air models. The next year it evolved into a separate model line. The SS (Super Sport) version first appeared in 1961 as the high-performance variant of the Impala lineup of convertibles, coupes, and sedans. The Impala SS name disappeared after the 1969 model year until it was reborn 25 years later as a high-performance four-door sedan. It was the first time a four-door carried the Impala SS logo, but it was a worthy successor. The reincarnated Impala SS was offered from 1994 to 1996, when its production ceased along with Chevy's commitment to large rear-drive cars.

The inspiration for the Impala SS arose from the popular trend among the hot rod set during the early 1990s of lowering and customizing the big Caprice Classic (1991–1996 B-Body) station wagons and sedans. The large size, combined with the controversial

The last of the big rear-wheel-drive American musclecars had a 260-horsepower, fuel-injected 5.7-liter LT1 V-8 under the hood. A four-speed automatic and limited-slip differential were also part of the package that borrowed heavily from the police-package catalog.

styling of a smooth-sided humpback body that tapered down at either end, had earned the Caprice Classic the nickname of "whale." When dechromed, lowered, and fitted with larger wheels and tires, these cars are transformed from a cetacean into a retro custom cruiser.

Chevrolet endorsed the customizing trend with a show car for the 1992 Specialty Equipment Market Association (SEMA) show in Las Vegas. Wearing monochromatic black paint and retro Impala badges along with 17-inch aluminum wheels, the Impala SS was a huge hit with the media and car enthusiasts, including Chevy's general manager at that time, Jim Perkins. On Valentine's Day 1994, 14 months after the SEMA show, the heartbeat of Chevy musclecar fans beat a little louder as the first Impala SS in 25 years rolled off the assembly line in Arlington, Texas.

From the outside, the Impala SS could easily be distinguished from its mundane Caprice Classic siblings by a paint job (available only in black in 1994) that extended the body color to the front and rear fascias, rocker moldings, door handles, key locks, grille, taillight moldings, and antenna base. A rear deck spoiler and Impala SS emblems replaced all other extraneous exterior ornamentation. The interior of the 1994 and 1995 models still retained most of the Caprice's dowdiness, including a column shifter, but bucket seats and a center console, complete with a floor-mounted shifter, were added in 1996.

What set the Impala SS apart was more than a fancy paint job. As Jake and Elwood Blues, noted aficionados of full-sized American performance sedans, would have said, it had a "cop motor" and a "cop suspension." Sitting under the hood was a 260-horsepower LT1 5.7-liter (350 cubic inches) V-8 straight out of the Caprice Classic Law Enforcement Package listing in the Chevy catalog. Similar engines could be found in the Corvette and Camaro Z28. A big part of the engine's performance was due to the latest in sequential fuel injection and electronic engine controls that allowed a 10.5:1 compression ratio. The police package also yielded the 4L60-E electronically controlled four-speed automatic transmission and 3.08 rear end with limited-slip differential.

1996 CHEVROLET IMPALA SS SPECIFICATIONS

Base Price	$24,905 MSRP, plus $590 destination
Engine	5.7-liter (350 ci) LT1 OHV V-8 with sequential fuel injection
Power Ratings	260 horsepower @ 5,000 rpm 330 ft-lbs torque @ 3,200 rpm
Transmission	Four-speed automatic with lockup torque converter; 3.08:1 rear end with limited-slip differential
Brakes	Four-wheel discs with ABS
Suspension	Front: Heavy-duty; independent with coil springs, deCarbon gas shocks and stabilizer bar Rear: Heavy-duty; four-link live axle with coil springs, deCarbon gas shocks and stabilizer bar
Wheels/Tires	8.5x17-inch cast-aluminum; P255/50 ZR17 radials
Wheelbase	115.9 inches
Length	214.1 inches
Curb Weight	4,221 pounds
Weight Distribution, Front/Rear	55/45
EPA Fuel Economy, City/Highway	17/26 miles per gallon

The big 4,200-pound bullet could hit 60 from a standing start in 7.1 seconds. *Motor Trend* ran an SS through the quarter-mile in 15.4 seconds at 91.1 miles per hour and proclaimed it significantly quicker than the archetype SS396 Impala of the musclecar period. Riding on 17x8.5-inch cast-aluminum wheels with P255/50 ZR17 BFGoodrich Comp T/A radials and a beefed-up police suspension with deCarbon gas pressure shock absorbers, the Impala SS could hold its own with imported sport sedans. Four-wheel disc brakes (Caprice Classic models had rear drums) with 12-inch vented rotors made stopping the SS as impressive as going around corners.

Chevy turned out 6,303 Impala SS models in 1994, and all of them were painted black. The low number of custom wheels available limited the first year's production of Impalas. In 1995, exterior color choices were broadened to include dark cherry metallic and dark gray-green metallic. Black was still the most popular color and showed up on almost half (9,858) of the 21,434 vehicles produced that year. There were 7,134 dark cherry cars and 4,442 in gray-green. Chevy officially halted produc-

Despite analog gauges, there was little to differentiate the interior of the 1994–1995 Impala SS from the more pedestrian Caprice Classic. In 1996, bucket seats, a tachometer, center console, and floor shifter were added, but by that time, Chevy needed the Impala production facilities to build more trucks.

tion of its rear-drive B-Body cars on December 13, 1996, although Impala SS production carried into January 1997 to meet a heavy demand that increased 1996 production to 41,941 units. Color-wise, there were 19,085 black, 12,180 cherry, and 10,676 gray-green Impalas. The total production for the Impala SS' short life as America's last rear-drive muscle sedan was 69,768. Many feel the car ran out of time well before customers. The Impala SS performance legacy shows that badge engineering, if done properly, can be a good thing for manufacturers and enthusiasts.

Camaro Z28

The Camaro Z28 is the oldest living survivor of the musclecar era. It debuted as a special high-performance option package in 1967. If GM sticks to its current plan, the Z28 will be relegated to the history books after the 2002 model year.

The 1967 Z28 performance package included a 290-horsepower, 302-ci V-8 engine; sport suspension; and 15-inch wheels and tires. It was Chevy's way to meet the 5.0-liter production engine limit imposed by the regulations of the SCCA Trans-Am series. In 1970, the Z28 engine size was increased to 350 cubic inches.

The fourth-generation Camaro debuted in 1993 with more than just a swoopy new body. The Z28 had a six-speed manual transmission and 275 horsepower. To celebrate, Camaro was chosen as the Indy 500 pace car and wore a special black-over-white paint job with white-painted wheels.

Chevrolet sold 633 pace car replicas. All were Z28s equipped with the $995 special paint and graphics package. Despite the all-American image, all Camaro production is now carried out at Ste. Therese, Quebec, Canada.

The Z28 option was discontinued after the 1974 model year. In mid-1977, the Z28 returned as a separate model in the Camaro lineup. Things stayed that way until 1988 when the Z28 was replaced by the IROC-Z model. This model was named to take full promotional advantage of Chevy's involvement with the International Race of Champions (IROC) race series. Specially prepared Camaros were used in the series that pitted professional race drivers from all forms of racing against each other. In 1991, after Chevrolet no longer was associated with the IROC race series, the Z28 model resurfaced as the IROC-Z disappeared along with the rights to use the series name.

The current Z28 model is built on the fourth-generation Camaro platform that was introduced in the 1993 model year. This latest version hit the streets with more than a swoopy new body going for it. Under the hood was a version of the LT1 small-

2001 CHEVROLET CAMARO Z28 SS SPECIFICATIONS

Engine	5.7-liter aluminum OHV V-8 with forced-air induction
Power Ratings	335 horsepower @ 5,200 rpm 350 ft-lbs torque @ 4,000 rpm
Transmission	Six-speed manual
Suspension	Front: Independent; upper and lower A-arms Rear: Live axle on torque arm and trailing links
Wheels/Tires	17x9-inch; P275/40 ZR17
Brakes	Four-wheels discs
Wheelbase	101.1 inches
Curb Weight	3,306 pounds
Length	193.5 inches

The big news for the 1993 Z28 was the aluminum-head LT1 5.7-liter engine it inherited from the Corvette. Although the Corvette's powerplant had a 300-horsepower rating, the Z28 engine was rated at 275 horsepower. The Camaro's engine was still good enough to propel the car from 0 to 60 miles per hour in 5.6 seconds; a good half-second faster than a 1993 Mustang Cobra.

The 1993 Z28 had dual air bags, the first on any Chevrolet. High-back cloth bucket seats were standard on all Camaros, but the Pace Car Replicas included special white-and-black upholstery.

A convertible was added to the Camaro lineup in 1994. Unlike the previous generation, all convertibles were factory-built alongside the coupes. The 1995 Z28 pictured was the first year the exterior mirrors were painted the body color instead of black. For a base price of $23,095, you received all the fresh air you could handle and 275 horsepower.

This 1997 SS Coupe features modifications by SLP Engineering, which were ordered through the Chevrolet dealer, that included a special hood with functioning scoop and forced induction, revised suspension, and 17x9-inch Corvette ZR-1 wheels.

First available in 1996 as a $3,999 option, the SS package bumped horsepower on the Z28 from 275 to 305. An optional free-flowing exhaust was good for 10 more horses.

block V-8 that first appeared in the 1992 Corvette. In the Z28, the aluminum-head 5.7-liter V-8 put out 275 horsepower (opposed to the 300 horsepower rating in the Corvette) and was mated to a new six-speed manual transmission. Also new was an unequal length A-arm front suspension. It was just what the Z28 needed to reclaim the pony car performance crown from the Mustang. *Motor Trend* put the 1993 275-horsepower Z28 head-to-head with a 1993 Mustang Cobra and got these results: 0 to 60 miles per hour, 5.6 seconds for the Camaro and 6.2 seconds for the Mustang; quarter-mile, 14.0 seconds (98 miles per hour) for the Camaro against the Mustang's 14.4 seconds (97.4 miles per hour).

In celebration of its racy new performance, the Z28 also served as the official 1993 Indy 500 Pace Car. Chevy made 633 pace car replicas that had the special two-tone black over Arctic White exterior, white-painted wheels, and Indy 500 graphics. The cost to outfit a Z28 Coupe ($16,779 base price including LT1 engine) as a pace car replica was $995.

In 1996, the *SS* initials returned to Z28s equipped with option R7T. These initials were accompanied by modifications that raised the performance bar for old and new Z28s. SLP Engineering carried out the R7T modifications with the approval of Chevrolet. They included a 305-horsepower version of the LT1 with a functional hood scoop and forced induction, reworked suspension, special rear spoiler, and 17x9-inch ZR-1–type wheels with P275/40 ZR17 tires. Camaro fans knew a good deal when they saw one. Despite the additional $3,999, 2,420 SS models were sold in 1996, followed by 3,430 more in 1997.

The Camaro under- went cosmetic surgery in 1998 with changes to the headlights, nose, grille, and rear taillights. Not everyone approved of the external changes, but the changes under the skin received a unanimous thumbs-up. The aluminum LS1 V-8 introduced as part of the all-new 1997 Corvette had filtered

In 1997, a 30th Anniversary appearance package was available that included an all-white exterior with white wheels and two orange stripes. The interior details were white leather trim with houndstooth seating surfaces.

One thousand 30th Anniversary Z28s were modified by SLP to SS specs. Part of the SS option was a new rear spoiler. After 1997, SS modifications were carried out at the GM assembly plant.

down to the Camaro. This new engine bumped the horsepower of a standard Z28 to 305 (later increased to 310 in 2001). *Motor Trend's* test of a 1999 Z28 came up with a 0-to-60 time of 5.5 seconds. When they tested a 2000 model Z28 with the SS option, now rated at 320 horsepower, the time was brought down to 5.3 seconds. On the drag strip, the SS turned in a quarter-mile elapsed time of 13.7 seconds—outstanding performance for a car priced around $30,000. The Camaro Z28 may be nearing the end of the trail, but it seems that Chevrolet has saved the best for last.

Pontiac Trans Am WS6

It is fitting that the Trans Am should remain a growling, menacing reminder of the bad ol' days when herds of musclecars prowled our streets and highways. Pontiac started the whole musclecar phenomenon in 1963 by stuffing a 326-ci V-8 inside a Tempest. The GTO followed in 1964, and the first Firebirds came out in February 1967.

The Trans Am started out as the WS4 option package in March 1969 on 1969 model year Firebird hardtops and convertibles. It consisted of the 400-ci HO engine (335-horsepower Ram Air III), three-speed manual transmission, heavy-duty suspension with stabilizer bars, Safe-T-Track rear end, a rear spoiler, special hood scoops, and a Cameo White exterior with blue stripes. Pontiac built 689 of these cars, and 55 of them had the optional 345-horsepower Ram Air IV version of the 400-ci V-8.

Since the beginning, Trans Ams have been heavy breathers. The latest WS6 iteration of the Trans Am is true to its "ram air" roots with two sets of hood-mounted scoops that, when coated in sinister black paint, take on the scowling visage of Darth Vader, in addition to his respiratory characteristics. The Trans Am has built a reputation of being one bad dude over the years, and this can also be seen in the transformation from the "good guy" white-only exterior in 1969 to the preference by latter-day owners for an all-black paint scheme.

The WS6 option code dates back to 1978, when it was first offered as a Trans Am Special Performance Package that consisted primarily of larger wheels and tires and a larger rear stabilizer bar. Although its content has changed over the years to include items including four-wheel disc brakes, the WS6 remained more or less a handling package. The WS6 nomenclature rejuvenated the Trans Am's performance image as well as its deep-breathing musclecar reputation in 1996.

The Ram Air logo on the twin hood nostrils lets the world know that this is the heavy-breathing version of the Trans Am; capable of obtaining 14-second quarter-miles and 5.7-second 0-to-60 sprints.

Pontiac's WS6 Performance and Handling Package debuted in 1996. For $2,995, you could relive the good old days of Ram Air induction for your Firebird while upgrading suspension to include 17x8-inch cast-aluminum wheels; stiffer shocks, springs, and bushings; plus 31-millimeter front, and 19-millimeter rear stabilizer bars.

The 1996 WS6 Performance and Handling Package, available on Formula and Trans Am Firebirds, used a twin-port hood scoop Ram Air system to bump horsepower of the Corvette-derived LT1 5.7-liter V-8 from 275 at 5,000 rpm to 305 at 5,400 rpm. The torque also increased from 325 foot-pounds at 2,400 rpm to 335 foot-pounds at 3,200 rpm. A *Motor Trend* road test of a WS6 recorded 0 to 60 miles per hour in 5.7 seconds and a flat 14 seconds through the quarter-mile with a trap speed of 101.9 miles per hour.

To put this in historical perspective, the *Motor Trend* archives show that the original 1969 Trans Am with a 335-horsepower engine turned in a

Taupe was one of only two interior color options in 1996. The other choice was Dark Pewter, which coordinated better with GM corporate gray on the dash and center console. Only 7,936 Trans Ams were sold in 1996.

0-to-60-mile-per-hour time of 6.5 seconds and a 14.9-second quarter-mile. A 1971 455-HO (335 horsepower)—motored Trans Am was able to reach 60 miles per hour in 5.4 seconds while running down the quarter-mile in 13.8 seconds at 105.6 miles per hour.

Other goodies in the 1996 WS6 package included a stiffer suspension with larger front/rear stabilizer bars, firmer transmission mounts, and 17x8-inch five-spoke cast-aluminum wheels with P275/40 ZR17 tires.

In 1998, the aluminum LS1 5.7-liter V-8 was installed in the Firebird to give it 305 horsepower as a starting point. The WS6 package built on that and used a new hood with two functional hood scoops placed directly over the Trans Am's standard nonfunctioning hood scoops for the Ram Air effect. Besides the firmer suspension parts, a free-flowing exhaust was also part of the package, although the 1998 models had a single outlet as

The Trans Am received some revised bodywork in 1998. The WS6 option included a new hood that added dual functioning air scoops above the standard nonfunctioning set.

Like the Camaro Z28, the Trans Am came equipped with the aluminum LS1 5.7-liter V-8 in 1998. Rated at 305 horsepower in stock format, the power shot up to 320 horsepower while the 0-to-60 time shrank to 5.1 seconds after SLP Engineering did some respiratory work through the application of the WS6 Ram Air package.

A free-breathing exhaust system was part of the WS6 package in 1998. A large single outlet identifies 1998 models. Beginning in 1999, smaller dual outlets were used.

opposed to the dual outlets on 1999 and later cars. The power output for the 1998 WS6 Trans Am jumped to 320 horsepower and 345 foot-pounds of torque. This increase dropped the 0-to-60-mile-per-hour time to an incredible 5.1 seconds.

Much of the credit for the improved performance of the WS6 package should be given to SLP Engineering of Troy, Michigan, whose president, Ed Hamburger, is a former drag racer who started modifying Firebirds in 1991. His original Firehawk impressed GM so much that his company has been allowed to work closely with Pontiac and Chevrolet in offering special performance and handling packages such as the Camaro SS and the Firehawk through GM dealers.

In 1999, the Trans Am celebrated its 30th anniversary with a series of 1,600 cars—1,065 coupes and 535 convertibles—that carried a special white exterior paint scheme with twin blue racing stripes. The interior included white leather seats with 30th Anniversary logos on the headrests. The convertibles had blue tops, and all of the cars had blue-tinted 17-inch wheels. This vehicle may have been the last hurrah for the Trans Am, which is scheduled to cease production after 2002.

Chevrolet 454SS Pickup

While GMC took to the technological high road by employing a Euro-style approach (turbocharging and all-wheel drive) to develop the

The 454SS pickup debuted in 1990 with brutish good looks and 240 horsepower. In 1991, Chevy decided to match its performance to its appearance. A new intake manifold and dual exhausts boosted power to 255 horsepower. A new electronically controlled four-speed automatic and 4.10 rear axle also boosted acceleration.

1991 CHEVROLET 454SS PICKUP SPECIFICATIONS

Body/Chassis	Regular Cab, 6.5-foot bed/ladder-type frame
Engine	7.4-liter V-8 with electronic fuel injection
Power Ratings	255 horsepower @ 4,000 rpm
	400 ft-lbs torque @ 2,400 rpm
Transmission	Four-speed electronic automatic
Suspension	Front: Independent, coil springs, gas shocks, stabilizer bar
	Rear: Semifloating rear axle, multileaf springs, gas shocks
Wheels/Tires	15x7-inch; P275/60 R15
Brakes	Disc/drum with rear ABS
Wheelbase	117.5 inches
Length	194.1 inches
EPA Fuel Economy, City/Highway	10/12 miles per gallon

Syclone, its high-performance pickup, Chevrolet chose to do things the old-fashioned, all-American way. Reasoning that there is no substitute for cubic inches, especially in writing ad copy aimed at the muscletruck crowd, they stuffed a big-block V-8 under the hood of a standard-sized pickup.

The main ingredients of the 1990 Chevrolet 454SS were a C1500 Series Regular Cab pickup with a 7.4-liter (454 cubic inches) V-8, automatic transmission, 15-inch chrome wheels with performance radial tires, and a fancy interior. The 454SS nomenclature may have recalled Chevy's lusty high-performance past, but the performance of the new 7.4-liter V-8 truck engine did not. In 1990, the 454SS had 230 horsepower tied to a three-speed automatic transmission to propel a 4,720-pound truck. Torque was a respectable 385 foot-pounds. The best *Motor Trend* testers could coax out of their 1990 454SS was a 0-to-60-mile-per-hour time of 7.8 seconds. At the drag strip, the big truck ran a quarter-mile in 16 seconds at 85.8 miles per hour. This was a bit of a letdown, based on expectations arising from the legendary high-performance alphanumeric badges on its flanks.

The intent of this big truck could not be criticized. With its Onyx Black monochromatic paint scheme set off by the stylish chrome wheels, red *SS* logos, and a Garnet Red cloth interior, the 454SS was a handsome vehicle. It was also a pretty good deal when you consider everything that was included in the $18,295 base price.

The 454SS may not have brought pavement-ripping performance, but it did provide a nicely equipped sport luxury truck. The mechanicals included the fuel-injected V-8, three-speed automatic transmission, a locking 3.73 rear differential, performance-handling package with front stabilizer bar, variable-ratio power steering, and coolers for the engine oil and transmission. Five 15x7-inch chrome wheels and P275/60 HR15 tires (including the spare) were also part of the package deal. Although the carrying capacity of the 6.5-foot cargo bed was reduced almost in half to a bit over 1,000 pounds by suspension and tires that were geared toward handling, not hauling, the 454SS still maintained a decent portion of its functionality.

The exterior trim was basically the high-end Silverado trim package with a blacked-out grille and other chrome items. The bumpers were painted black. A front air dam included fog lamps, and halogen headlights and a sliding rear window were also part of the package.

The Silverado décor was carried over to the interior of the 454SS. Special upgrades were the Garnet Red luxury cloth high-back reclining bucket seats and center console. A four-spoke sport steering wheel

For 1992, red and white joined black as exterior color choices. Despite a high level of interior luxury, first-year sales of almost 14,000 units, and decent performance at a price tag right around $20,000, demand plummeted to less than 1,000 vehicles in 1993, forcing Chevy to pull the plug.

with tilt; tachometer; cruise control; air conditioning; intermittent wipers; power locks, doors, and windows added to the luxurious cabin. Rounding out the high level of creature comforts was a high-end stereo system including an electronically tuned AM/FM radio with cassette player and graphic equalizer. The 454SS was also the official pace truck of the 1990 Indianapolis 500. A decal set commemorating this feature was available.

In 1991, Chevrolet made significant changes to the 454SS powertrain that upgraded its performance to match its appearance. The 7.4-liter V-8 received a new one-piece intake manifold with a relocated throttle body injector. A dual exhaust system helped boost power output to 255 horsepower at 4,000 rpm, and 400 foot-pounds of torque at 2,400 rpm. The 4L80-E heavy-duty electronic four-speed automatic transmission was also new. This "smart" transmission was able to compensate for variations in temperature, altitude, and engine performance to enhance performance, reliability, and fuel consumption. Straight-line performance was also improved, and a 4.10:1 rear axle and Bilstein gas shocks also became part of the 454SS package. For 1992, the 454SS was available in red and white exterior colors, as well as black.

When *Motor Trend* tested a 1993 454SS, the price had risen to $21,240, but the level of performance had also increased. The 0-to-60 time had come down to a respectable 7.2 seconds, and the quarter-mile was reached in 15.8 seconds at 84.7 miles per hour. This was the last hurrah for the muscular truck, which was not offered the next year. Despite strong sales of almost 14,000 units in 1990, Chevy was only able to sell a little more than 3,000 total units over the next three model years. So for 1994, the C1500 pickup went back to its day job as a work truck and the 454SS nomenclature was returned to Chevy's hall of fame.

GMC Syclone

Like a real cyclone, the GMC Syclone suddenly stormed onto the sport-truck scene, kicking up a lot of dust accompanied by the wail of its turbocharged V-6, and then just as quickly it was gone. GM built

GMC

Previous page: *Despite its boxy looks, the 1991 GMC Syclone could do 0 to 60 miles per hour in under 5 seconds and reach a governed speed limit of 126 miles per hour.*

only 2,995 Syclones ("*S*" indicating the S15 pickup body on which it was based) from January through July 1991. Despite all the sporty pickups that have come and gone since then, no one has built a faster truck. Although the body configuration required the top speed to be limited to 126 miles per hour, acceleration tests by *Motor Trend* show the Syclone was capable of 0 to 60 miles per hour in 4.9 seconds with a quarter-mile time of 13.6 seconds at 98.6 miles per hour.

Motivating the Syclone was a turbocharged, beefed-up version of the 4.3-liter Vortec V-6 found

The Syclone may have looked like an S15 pickup, but underneath it had the heart of a 959 Porsche with all-wheel drive, turbocharged engine, ABS braking at all four corners, and 16-inch aluminum wheels. The base price was $25,970.

1991 GMC SYCLONE SPECIFICATIONS

Body	Regular Cab, Shortbed pickup
Engine	4.3-liter V-6 with turbocharger, intercooler, and multiport fuel injection
Power Ratings	280 horsepower @ 4,200 rpm 360 ft-lbs torque @ 3,600 rpm
Transmission	Four-speed automatic with overdrive; all-wheel drive
Suspension	Front: Independent, torsion bars Rear: Semi-elliptic, two-stage multileaf springs
Wheels/Tires	16x8-inch aluminum; P245/50 VR16
Brakes	Front disc/rear drum with four-wheel ABS
Curb Weight	3,613 pounds
Wheelbase	108.3 inches
Length	180.5 inches
EPA Fuel Economy, City/Highway	14/17 miles per gallon

in more mundane GMC pickups. The engine featured a compression ratio of 8.35:1 and an electronically controlled, multiport fuel injection. This was the dawning of the electronics age for engine management systems. By using this system, GM was able to minimize turbo lag. A liquid-cooled intercooler assisted in getting horsepower up to 280 at 4,400 rpm. The torque was equally as impressive with 350 foot-pounds at 3,600 rpm. Handling all this power was a recalibrated version of GM's TH700-R4 four-speed automatic with overdrive as used in the Corvette. The final drive ratio was 3.42.

GM's high-tech approach to building its sport pickup continued when all that turbo power was delivered by the automatic transmission. No squirrelly rear-wheel driving antics for this powerful pickup.

The 4.3-liter V-6 with turbocharger, liquid-cooled intercooler, and state-of-the-art electronically controlled multiport fuel injection was very sophisticated for 1991. The engine produced 280 horsepower and an even more impressive 360 foot-pounds of torque.

The Syclone interior featured upgrades such as special cloth bucket seats with lumbar adjustment, leather-wrapped sport steering wheel, air conditioning, cruise control, analog instrumentation including turbo boost gauge, console with cup holders, and an electronically tuned AM/FM/cassette stereo with seek/scan, digital clock, and graphic equalizer. GMC sold 2,995 Syclones in 1991.

The Syclone had full-time, all-wheel drive. It also had a front disc/rear drum brake setup that included ABS at all four corners—a rarity even in cars at the time, let alone trucks. While most sport pickups tend to emulate musclecars, the Syclone was trying to be a Porsche 959 with a cargo bed.

Other refinements included with the Syclone were aluminum 16x8-inch wheels wrapped in P245/50 VR16 performance tires. The S15 Regular Cab pickup body was draped in a monochromatic black paint scheme and received special trim and an aerodynamic body kit, including a removable tonneau cover.

The interior upgrades consisted of special cloth bucket seats with lumbar adjustment, a leather-wrapped sport steering wheel, air conditioning, cruise control, analog instrumentation including a 120-mile-per-hour speedometer and turbo boost gauge, floor shifter, console with cup holders, and an electronically tuned AM/FM/cassette stereo with seek/scan, digital clock, and graphic equalizer. Remember, this was 1991 and the latter was just about as high tech as a factory sound system could be at the time.

Marlboro had 10 Syclones customized as part of an advertising sweepstakes promotion. These vehicles were painted PPG Hot Lick Red and had T-top roofs, Recaro seats with five-point racing harnesses, a power drop-down rear window, and a CD player.

If you were not lucky enough to win a Syclone, the base price was about $25,970. Because of the changes to the suspension and the all-wheel-drive system, the cargo box load rating of a Syclone was 500 pounds. Despite the bang for the buck it delivered, the limited carrying capacity of the Syclone was probably a contributing factor to its short production life.

To remedy this situation, GMC introduced the Typhoon, which was essentially the running gear of the Syclone mounted in a two-door Jimmy SUV body, in 1992. The Typhoon engine produced 5 more horsepower (285 horsepower) than the Syclone. The Typhoon was also geared to a more upscale crowd, with a luxurious interior including leather seats. Unlike the basic black of the Syclone, Typhoon buyers were able to choose from a wider color palette. The covered body effectively doubled the Syclone's carrying capacity for people (up to five) and cargo (900 pounds). It also cost more, with sticker prices hovering around the $29,500 range. The Typhoon weighed about 250 pounds more than the Syclone, so its performance was slightly slower, but still impressive. The Typhoon was offered in 1992 and 1993. There were 2,500 1992 models produced, and 2,200 in 1993.

At the time of its debut, the Syclone may have been the answer to a question no one had thought of asking. After all, back then the phrase "sophisticated truck" was an oxymoron. Now that light trucks make up around half of new car sales a decade later, maybe the climate is right for the Syclone to hit again.

Index

Other MBI Publishing Company titles of interest:

Corvette C5
ISBN: 0-7603-0457-2

Camaro
ISBN: 0-7603-0783-0

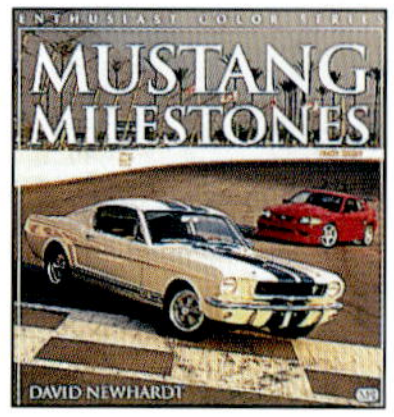

Mustang Milestones
ISBN: 0-7603-0971-X

Hemi Muscle Cars
ISBN: 0-7603-0635-4

Mustang 1964 1/2-1973
ISBN: 0-7603-0734-2

Corvette 1968-1982
ISBN: 0-7603-0418-1

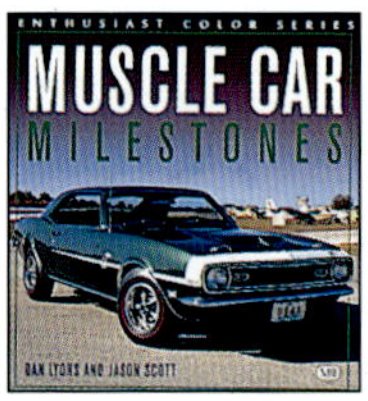

Muscle Car Milestones
ISBN: 0-7603-0615-X

Dodge Viper
ISBN: 0-7603-0984-1

Mustang 5.0 & 4.6
1979-1998
ISBN: 0-7603-0334-7

Find us on the internet at www.motorbooks.com 1-800-826-6600